Electric Machines in Agriculture

Electric Machines in Agriculture:
Origins, Developments and Applications

Kevin Desmond
Bordeaux, France

NEW INDIA PUBLISHING AGENCY

101, Vikas Surya Plaza, CU Block, LSC Market
Pitam Pura, New Delhi – 110 034, India
Email: info@nipabooks.com
Web: www.nipabooks.com

For customer assistance, please contact

Phone: + 91-11-27 34 17 17 Fax: + 91-11- 27 34 16 16
E-Mail: feedbacks@nipabooks.com

ISBN: 978-93-89547-96-2

Composed and Designed by NIPA.

M.S. SWAMINATHAN RESEARCH FOUNDATION

M.S. Swaminathan
Founder Chairman
Ex-Member of Parliament (Rajya Sabha)

Foreword

In the search for Biohappiness, there is one element which can make a vital contribution to the Evergreen Revolution, both in India and around the world. This is the intelligent use of electric agricultural machinery. To feed an exploding population, precision farming using sustainable energy sources will play a crucial part Kevin Desmond's book not only shows how electricity has gone hand in hand with mechanised farming from as far back as the 18th Century. It also presents an impressive arsenal of machinery available today and in the coming years which may enable us to achieve that imperative goal.

M S Swaminathan

M. S. Swaminathan

M.S. Swaminathan is an Indian geneticist and administrator, known for his role in India's Green Revolution, which transformed India's image then as begging bowl to bread basket. This transformation during the 1960s took just about 4 years. The yield increases achieved in wheat and then in rice which occurred in just about half decade is far in excess of the yield increases during the preceding 4000 years.

Acknowledgements

The author would like to thank the following people and organisations for their help in preparing this history:

Archives Départementales Pyrenées Orientales; Archives Départementales de la Marne; Caroline Benson (Museum of English Rural Life, University of Reading); Ewald Blocher (Siemens); Richard Borlase Matthews; Cathays Branch and Heritage Library (Cardiff); Stephonie J Clarke (Archivist for the Felbridge History Group); "Cragside" House (Andrew Sawyer, Conservation Manager); Stephen Heckeroth; Ciccinella Kechler; Ted Kemp (IEE); Damien Kuntz (Musée EDF, Electropolis); Jean-Noël Raymon; the Small Robot Company; Salah Sukkarieh (Australian Centre for Field Robotics (ACFR);

Thanks also to Sumit Pal Jain and the team at NIPA, to my wife Alexandra Desmond (my long-supporting wife) and Kathryn Cooper (my indexer).

Preface

Rather like the acronym UAVs is used to describe Unmanned Aerial Vehicles, or LEVS to describe Light Electric Vehicles, or PDDs to describe Personal Delivery Devices, I should like to propose a new acronym, EAMs to describe Electrical Agricultural Machines and so EAMS will be used to describe all kinds of electrical machinery used in farming from early tractors to the drone.

EAMs, which can be for static use or for transportation, can be traced back over two hundred and seventy years ago to the year 1746. Alessandro Volta was an Italian physicist, chemist, and a pioneer of electricity and power, who is credited as the inventor of the electric battery. At the very young age of 7, Alessandro started to show a great interest in natural phenomena, to the extent that, anxious to find some bright blades of straw, which according to the local farmers were supposed to be made of gold, he almost drowned in the Monteverde spring, near Camnago. He did not find it, fifty-three years later, he would invent the fundamental building block for electrical power: the Voltaic pile.

Despite a century of horrific wars, and Industrial Revolution and some significant international plagues, human populations managed to grow 400 percent during the 20th century. Today's world population is 7.6 billion, and the United Nations projects that another billion will be added by 2030, still another by 2050, and that by 2100, the world population will be 11.2 billion. The food they will eat will need to be farmed as efficiently and as sustainably as possible. It is therefore vital that this book looks at EAMs and their tremendous potential to deal with this.

Kevin Desmond
Bordeaux, France

Contents

1

Precursors (1850-1914)

The domestication of oxen in Mesopotamia and the Indus Valley Civilization, perhaps as early as the 6th millennium B.C. provided mankind with the draft power necessary to develop the early plough into the larger, animal-drawn true ard (or scratch plough). Some of the earliest evidence of ploughing from around 2800 B.C.was found at the Indus Valley site of Kalibangan. Subsequent archaeological finds in Prague, Czech Republic, push the oldest known ploughed field even further back to 3500-3800 B.C. All food produce was planted and harvested by hand: seeds were scattered by hand and planted grains were brought from the field to be threshed on the dooryard. Milking a small herd of cows or goats was also done by hand. This continued for centuries. Then man discovered electricity.

In 1746, just a few years before Benjamin Franklin of Philadelphia sent a kite to catch lightning in a storm, a Doctor von Maimbray gave talk to the Royal Society of London where he described his experiment of electrifying the myrtle bush (Morella cerifera) which then put forth new branches in October, something which had never happened before. This news inspired others in Europe to carry out similar experiments. In 1749, Abbot Jean Antoine Nollet of the Royal Academy of Sciences in Paris France, inventor of the electroscope, published a pamphlet on "The Harmful Effects or Advantageous of Electrical Phenomena on Plants."[1]Isolated experiments continued until in 1783 Father Pierre Bertholon de Saint-Lazare of Montpellier published "About the Electricity of Plants", a work in which he dealt with the effect of atmospheric electricity on plants, taking in their medicinal and nutritive-electrical virtues and but in particular the practical means of applying it usefully to agriculture.[2]He even built an electro-vegetometer, a device for collecting atmospheric electricity to distribute it in the soil and then to the plants, but he met with little success. Bertholon also conceived of subjecting plants to "electric rain" by connecting a large syringe to an electric source. A gardener, standing on a slab of insulating material in a wagon, carried a sprinkling can electrified by means of a wire leading to a static machine. "By means of this process, strange for the times, the good Abbé Bertholon, who was considered something of a sorcerer, obtained lettuces of an extraordinary size."[3]

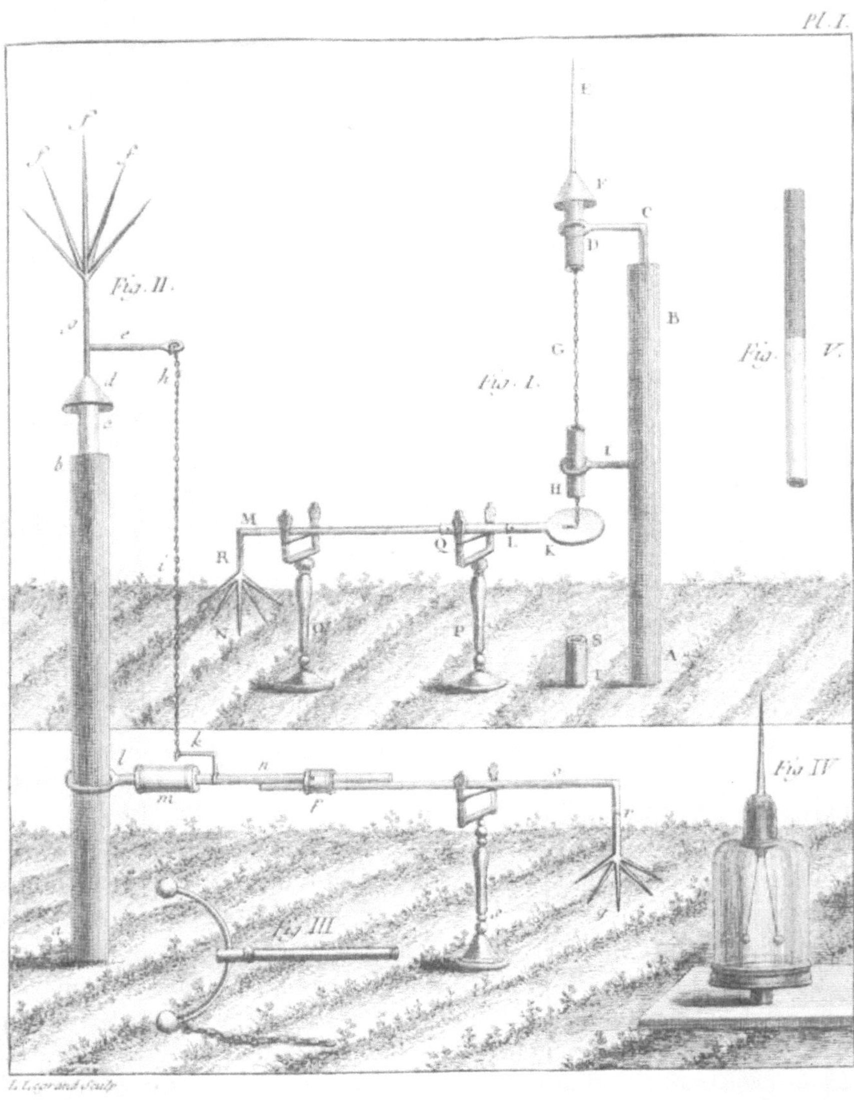

Following the work of Maimbray in Scotland and Nollet in France, in 1783 Father Pierre Bertholon de Saint-Lazare of Montpellier invented this electro-vegetometer. It would "extract" electricity from the sky, from a top of a mast, to lead to the ground and to increase plant growth. This world's first electric agricultural machine proved not very successful. This copper plate print is signed "L. Legrand, sculp." (Musée EDF Electropolis, Mulhouse).

Following the work of Maimbray in Scotland and Nollet in France, in 1783 Father Pierre Bertholon de Saint-Lazare of Montpellier invented this electro-vegetometer. It would "extract" electricity from the sky, from a top of a mast, to lead to the ground and to increase plant growth. This world's first electric agricultural machine proved not very successful. This copper plate print is signed "L. Legrand, sculp." (Musée EDF Electropolis, Mulhouse).

While experiments were made with primitive electric motors, it was at first the steam engine which was harnessed into the service of agriculture. In 1832, John Heathcote, director of a lace-making factory in Tiverton, Devon took out a patent for a steam plough. His factory had a foundry for making horse-drawn ploughs so before long a prototype was constructed. Two continuous tracks of heavy canvas with wooden lags went around wheels 8 feetin diameter. On a low platform between the wheels were the steam boiler and engine, as well as two winding drums at right angles to the tracks. Steel cables were not yet available, so each drum was wrapped by a long, flexible hauling band made of iron strips fastened together.The engine ran on a raised, rolled roadway constructed through the middle of a field. A haulage band was run to each side from a winding drum across the field, around a pulley on a movable anchor cart and back to the other winding drum. A single-bottom walking plough worked on each side. When one plough was going out, the other was coming back to the engine. At the end of each round, the two anchor carts and the engine were moved ahead.Following a rather clumsy demonstration in Scotland, Heathcoat gave up his project having invested £12,000 in the machine. (About £1.3 million or $1.5 million today.) Unknown to Heathcote, at the same time, Edmund C. Bellinger of Barnwell, South Carolina,USA had sketched out a plan for a cable-operated steam plough. Imaginative machines were produced by Major Amos Tyrell of Genesee County, New York, and the distinguished naturalist, Professor Constantine Rafinesque of Philadelphia, but neither proved successful. In the decade of the 'forties the ponderous contraptions of Larkin and Cowling caught the public eye but rumbled into obscurity as rapidly as they had appeared.

Another prototype steam tractor was commissioned by Peter Robert Drummond Burrell, Baron Willoughby de Eresby, on his estate near Grimsthorpe, England. This 26-horsepower agricultural locomotive weighed some three tons. It was manufactured in 1850 especially for his lordship, and designed by Mr. Gooch, the locomotive superintendent of the London and North Western Railway. The tractor moved on a portable railway, and the ploughs advanced and receded on either side of it at right angles, precisely as in the case of John Heathcote's engine, and it was followed by as many subsoil ploughs, so that both operations were performed at once.

A third prototype was built by Tullocks for the Marquis of Tweedale in 1857. It used two engines. The front wheels could be turned and the drive was provided from the rear wheels. Two ploughing speeds, 3½ and 5 mph (5.6 and 8 kph, were provided. The Marquis thought it necessary to clear the land of flints and stubble before ploughing, so invented his horse-drawn subsoil plough. When the steam plough came within a certain distance of the engine it was lifted clear of the ground by means of a cranked shaft attached to the horizontal beam above the engine. This shaft was moved by engine power and the plough turned over on its axis. Another turn of the crankshaft lowered the plough and positioned it for its return run across the field. Whilst the plough was moving away the engine moved forward 56 inches 142 cm which was the width of four 14-inch 36-cm furrows. When moving on the highway the top fixtures and all the accessories were carried on a wagon at the rear of the engine. Fully laden the machine could attain a speed of 4 mph 6.4 kph.

In the USA in 1855, Obed Hussey, better known for his reaper design patented in 1833, demonstrated his own steam plough. Hussey worked on the design of his plough for at least a few years, but failed to bring it to market. Richard Gatling, the inventor of the Gatling Gun, worked on the development of a workable steam plough, receiving a patent in 1857.

Joseph Walker Fawkes, the son of farmerin Lancaster County, Pennsylvania, USA, had received a limited education and had been apprenticed to learn the carpenter's trade. Within a few years he had become the proprietor of a small machine shop at Christiana, the town of his birth. A mechanic of considerable innate talent, he gradually became more and more aware of farm problems and requirements in a world on which the impact of the Industrial Revolution was becoming ever greater. His travels through the prairie-land of the Mid-West ultimately set him to work on the application of steam to tilling the soil. By the autumn of 1855 he had fashioned a model of a steam-driven ploughing apparatus which provided traction through a large driving drum, rather than through conventional wheels. To a friend, J. G. Dickinson, Fawkes took his model, carefully tied up in a handkerchief, and asked for financial aid. Dickinson and several associates contributed and the inventor turned to the construction of a full-scale working model. But obstacles were many: the first engine was an utter failure and never left the shop. The second lacked the power necessary for traction and could not be used. Finally, in 1858, on the third attempt, the steam plough was ready for testing and a patent was granted to Fawkes for an "Improvement in machines for Ploughing." As described by contemporaries, the engine and ploughing gear attached to its rear were eighteen feet long. Its main frame was of iron, 8 feet wide by 12 feet long (2.4 m by 2.7 m), resting on the axle of a roller (driver), 6 feet (1.8 m) in diameter and 6 feet wide. One cylinder, 9 inches (23cm) in diameter with 15-inch(38-cm) stroke was provided on each side of the boiler. The first public trials of the Fawkes rig seem to have been in Decatur at the Lancaster County Fair in 1858, where it attracted considerable attention. Meanwhile, the Illinois State Agricultural Society, in conjunction with the Illinois Central Railroad, had offered premiums totalling $5,000 for the best steam engine adapted for ploughing and other farm work. In hopes of winning development capital, Fawkes took his machine to the Illinois

State Fair at Centralia in mid-September where trials were to be held before a select committee of the Society. There the ten-ton behemoth was the darling of the show, as exhibits of horse-drawn equipment were forgotten or ignored. The inventor became the man of the hour and the Chicago Press and Tribune slyly suspected that he was a descendant of Guy Fawkes, "as he is determined to blow up and out of existence the whole system of horse, mule and ox-ploughing, and substitute therefore his pet nag made of iron, whose breath is fire, and whose food is the product of diluvian and post diluvian forests." At the actual trials, the farmer upon whose field the running tests were to be made refused to allow the snorting iron monster on his land, and hard, baked, unbroken prairie was substituted. Fair-goers gathered at the site and only the noise of the engine drowned out the buzz of enthusiasm from onlookers as the approached. The following year Fawkes returned to Illinois with a new model, the *Lancaster*. He went to Moline, bought eight John Deere ploughs, bolted them together and used them in his winning demonstration in Chicago against James Waters of Detroit and John Van Doren of Chicago in the famous Illinois State Agricultural Society contest of 1859.That same year, President Abraham Lincoln addressed the Wisconsin State Agricultural Society at the Wisconsin State Fair in Milwaukee. Lincoln talked about the steam plough, about what he thought it should be like and the impact on American agriculture that would follow its development. In an oft-quoted speech, Lincoln told the assembled crowd that:

The successful application of steam-power to farm work is a desideratum - especially a steam plough. It is not enough that a machine operated by steam will really plow. To be successful, it must, all things considered, plow better than can be done by animal power. It must do all the work as well, and cheaper, or more rapidly, so as to get through more perfectly in season; or in some way afford an advantage over plowing with animals, else it is no success."[4]

The first public trials of the Fawkes steam plough seem to have been in Decatur at the Lancaster County Fair in 1858, where it attracted considerable attention.

The Civil War in America, which lasted from 1861 to 1865,where ploughs were converted to weapons, put pay to steam-powered development in that divided country.

Back in the England of Her Majesty Queen Victoria, in 1854 the Royal Agricultural Society of England (RASE) offered a prize of £200 for "the steam cultivator which shall in the most efficient manner turn over the soil and be an economical substitute for the plough or spade." One of those who competed was John Fowler of Wiltshire.

Between 1850 and 1864 Fowler took out in his own name and in partnership with others thirty-two patents for ploughs and ploughing apparatus, reaping machines, seed drills, traction engines, slide valves, the laying of electric telegraph cables, and the making of bricks and tiles. Fowler had already designed and built a series of steam-driven drainage ploughs. But for this machine he developed a mechanism that enabled it to plough in either direction without having to be turned round. Fowler achieved this by designing a frame that had two ploughs attached as a kind of see-saw. One of the two plough blades would be swung down to make contact with the soil, depending on the direction the plough was to travel. At the end of each furrow the anchored pulleys would be moved slightly ready for the next furrow. The firm of Ransome and Sims Ltd. built the new engine at its Orwell works at Ipswich, and on 10th April, 1856 a trial was carried out at Nacton in which one acre was ploughed in an hour. Despite the fact that the engine and plough coped well with the task, the effort of re-positioning the pulleys at either end of the field was too time-consuming. Although Fowler demonstrated his roundabout system at the Royal Agricultural Society of England meeting at Carlisle in 1856, he did not win the prize.

Fowler next using a weighted cart with a pulley mounted beneath the frame. The cart had disc wheels that dug into the ground so that the cart acted as an anchor for the pulley. Two carts would be placed at opposite ends of the furrow so as to pull the plough in either direction, and after completing a furrow, the carts would be winched to the position for the next furrow. He presented this modified version at the RASE Show in Chelmsford in 1857 where the prize had been raised to £500. His system was pitted against a rival ploughing system designed by John Smith of Woolston. Smith's design did not fulfil all of the judges' conditions and it was disqualified. Fowler's system worked very well, but the estimated cost of his work was 7s 2½d per acre as against 7s for horse ploughing. The prize was withheld. A similar trial was held at the R.A.S.E. meeting at Salisbury in 1857, but again the prize was withheld. The judges however awarded a medal to John Fowler as a reward for his strenuous endeavours, adding that: "steam ploughing as such had attained a degree of excellence comparable in point of execution with the best horse work."This was a bitter disappointment for Fowler who thought that the superior speed of his system over horse ploughing should have been taken into account. However, he did receive £200 awarded by the Royal Highland and Agricultural Society of Scotland, after a trial at Stirling that same year, despite the judges agreeing that the sole entrant had not exactly fulfilled the conditions the efforts were impressive.[4] Fowler returned to contest the R.A.S.E. ploughing trial at Chester in 1858. He was opposed by a number of competitors but was successful in being awarded the £500 prize.[5]One of the first clients for his plough was Viscount de Baulny, who bought one for use on his lands at Charny and Villeroy. Later he presented it to the Emperor Napoleon III, the latter using it to create a model farm surrounded by a forest at Vincennes.

By the 1860s, John Fowler's steam-powered cable ploughs were being put to work in farms across England and abroad. (The Museum of English Rural Life, University of Reading)

By 1858 Fowler had forty sets of ploughing tackle in use, and by 1861 he had one hundred sets working. From 1860 the manufacture of the ploughing machinery was carried out by the firm of Kitson and Hewitson of Leeds. Fowler's ploughing sets were sold all over the world and were responsible for bringing land into production that was previously unable to be cultivated.In 1873, the Pasha of Egypt employed 400 steam ploughs on his extensive cotton fields. Fowler's biographer, McCutchan documented over 3,000 single and double engine sets made between 1860-1920, including exports to Egypt, India, Australia, Argentina and Cuba.Sadly, John Fowler did not live to see the success of his machine as he died in a hunting accident in 1864. The Fowler steam plough system had a virtual monopoly, with improvements, until the last ploughing engine left Fowler's works in England in 1933, superseded by the development of the internal combustion engined tractor.[6]

Alongside steam power there was hydroelectric power. When, aged 68, retired industrialist and scientist William George Armstrong built his mansion, "Cragside" in Northumberland, he installed the world's first hydroelectric power system. Creating the Debdon Lake, Armstrong used its Tumbledown cascade to give power to a 12.5 kW dynamo built for him by his friend Carl Wilhelm Siemens and housed in a special outhouse. One evening in late November 1878, the library in "Cragside" was lit by a single electric arc lamp, the first domestic lighting in the world. Armstrong then extended the power to include more lighting, heating and power devices indoors and outdoors.

Electricity as a power source had remained limited to laboratory demonstrations until 1859 when Gaston Planté of Meudon, near Paris succeeded in developing the world's first rechargeable lead-acid battery, which would soon herald the dawn of battery-powered machines. Early applications of electricity however, were of necessity restricted to power and some lighting, although the full value of lighting was not completely realized for years.

Over in France's Third Republic one of its citizens, Auguste Bernard Albaret of Liancourt in France's Oise Region, took an innovative approach to his agriculture. He had won medals for the ingenious applications of his 5hp locomotive to grain threshing, straw baling and root-cutting. In 1878, Albaret applied steam power to Gramme electric motors enabling him to light up a field at Mormantin the department of Seine-et-Marne, and carry out night-time harvesting.

Fowler steam engines provided electric power to light up electric harvesting ((Musée EDF Electropolis, Mulhouse).

The Albaret system was made up of an ordinary locomotive producing the driving force, one or more electric motors, a bracket to carry the lantern and the regulator, all mounted on four wheels in order to make the movement of the apparatus easy. Albaret's workshops were also lit by electric light. As one journalist wrote:

This lighting system will doubtless soon by adopted in a certain number of agricultural factories.[7]

The motors used by Albaret had been purchased from Zénobe Gramme. During the early 1870s, Zénobe Théophile Gramme, a Belgian electrical engineer, perfected a type of direct current dynamo capable of generating smoother and much higher voltages than the dynamos known to that point. In 1873, Gramme accidentally discovered that the device was reversible and would spin when connected to any DC power supply. Gramme and his French business manager Hippolyte Fontaine took their motor to the Vienna World Exhibition, where they demonstrated the reversibility of the electrical generator and the transmission of electricity over a 1.24 mile (2km) distance via copper wiring. Before Gramme's inventions, electric motors attained only low power and were mainly used as toys or laboratory curiosities; the Gramme machine was the first usefully powerful electrical motor that was successful industrially.

An electric tractor presented a bigger challenge. In the early 1870s French engineer, Jean Chrétien, had invented several types of steam and direct-lift equipment, including a steam crane that had a higher efficiency (80%) over competing steam cranes (40%) and replaced hydraulic cranes (22% yield), which were then the most commonly used.

In 1875, Clément Félix bought ten 500 franc shares ina sugar factory and distillery in Sermaize les Bains in the Haute Marne region. He rented 1,000 acres (400 hectares) of land at the Tournay farm, close to Favresse where, with his manager Henri Marois brought from the Dordogne, they began to apply modern methods of cultivation, equipping the farm with steam ploughs. They added a creamery and built a railway line between Tournay and the port of the Brusson Canal to take the crops and load them onto barges destined for Sermaize. Félix even opened a bakery at nearby Brusson.

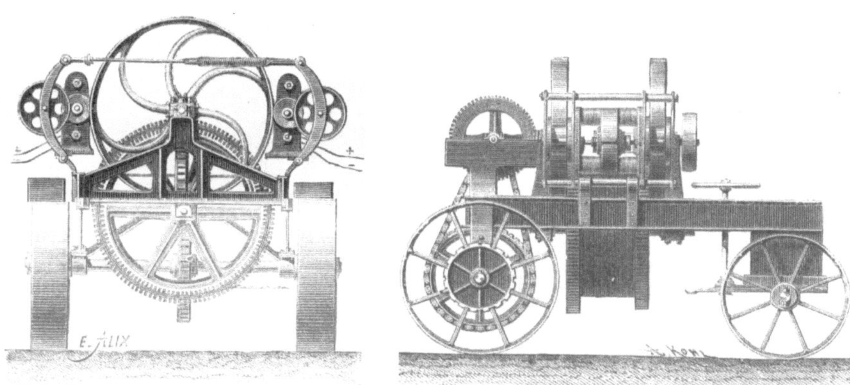

Félix Chrétien's ingenious coupling of a winch system to the electric motor developed by Zénobe Gramme. (Musée EDF Electropolis, Mulhouse).

Like Albaret, Clément Félix saw the advantages of the Gramme motor for agriculture, and working with Jean Chrétien, heingeniously linked two motors to a copper cable whose electricity came from his factory. On Thursday 22nd May, 1879 (the Roman Catholic Ascension Day holiday), he organised a demonstration of "labourage électrique" or electric ploughing of a field of beetroots at Sermaize, in the presence of delegations from a commission of the National Society of Farmers, of the French Agricultural Society as well as the local Prefect and the Commander-in-Chief of the 6th Army Corps. Henri Tresca, director of the Conservatoire des Arts et Metiers de Paris had arrived from the metropolis with a group of interested electrical engineers. According to the press, 1,200 to 1,500 people from the towns and villages of the region came to witness the trials[8].

At each end of the field to be worked, they had placed a pulley which commanded two machines. The current was run by two conductors of about 328 yards300 metres length, the plough was attached to the pulley by steel cables which gave it a to-and-fro movement by successively coiling and uncoiling. The electric plough took about one minute to work 215 ft^2 (20 m^2) of land, which corresponded to 5-6 steam horsepower. The cables were then moved along until the field was finished.[9] The end of the trial met with the unison clapping applause of the onlookers.[10] In terms of electrical supply, it was explained that a farmer might also harness, rivers and waterways and canal lock gates or windmills or in one word, Nature.

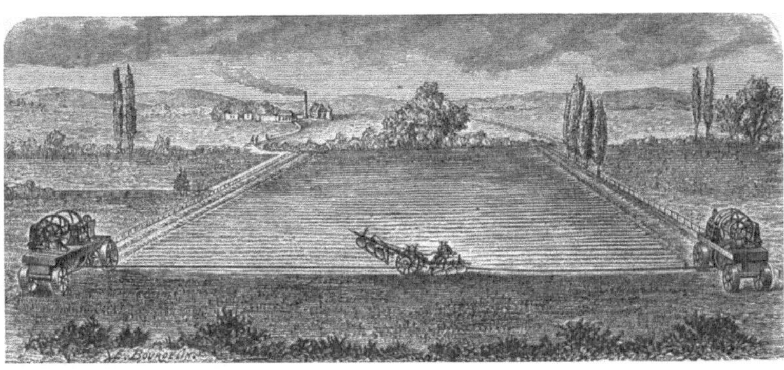

The beetroot field at Sermaise-les-Bains was the first to be ploughed using a cable-electric system.

This was the birth of electric agricultural machinery.

During the 1878-1879 season, 400 tons (363 tonnes) of beetroot were taken the 100 yards (91 metres) to the port of Sermaize where they were loaded onto barges for a journey up the canal de la Marne au Rhin.

In September, 1878 Clément Félix also lit up the music festival evenings organised by the local brass band. Using the same Gramme motors, six electric Jablochkoff candles cast bright light across the fete. A powerful reflector placed at the top of the sugar factory projected light for more than 550 yards (500 metres). For one evening, the little town of Sermaize was lit up like the capital city of Paris!

In May, 1880 Félix made a second experiment in electric ploughing at the Bar-le-Duc Agricultural Show in front of a special jury and watched by the Minister of Agriculture and Public Works. It ploughed 370 acres (150 hectares) and earned Félix a gold medal from the French Agricultural Society. The same demonstration was planned in Reims during the congress of the French Association for Scientific Progress. In August and September, 1881 Félix presented his electric plough, alongside a Gramme-engined wood-cutting saw at the very well attended International Electrical Exhibition in Paris[11], not only earning him a gold medal but also leading to his being made a Chevalier of the Legion of Honour (France's highest award) on 29th December, 1881. This decoration was presented to him on 14th June, 1882 by Hippolyte Fontaine, a civil engineer but also administrator of the Sucrerie-Distillerie de Sermaize. Félix was also elected Mayor of Sermaize, and during the following two decades, he developed a thriving sugar-refining business in the town[12].

Perhaps inspired, others were thinking along similar lines. The chocolate establishments of Menier in Noisiel, Seine-et-Marne, also applied to their property this Gramme electrical system, supplied to the factory by steam engines or hydraulic motors driven by the flow of the Marne.

Over in the U.S.A, an early use of hydroelectric energy for agricultural produce was made in Minneapolis, Minnesota, known as "Mill City" for its flour production. In 1879, after five years of secret planning, Charles Alfred Pillsbury announced to the public that he would build the largest and most advanced flour mill the world had ever seen. He had travelled to mills all over the world, searching for the best technique for milling flour on a large scale. This included a visit to Baron Armstrong's dynamo at "Cragside", Northumberland. From 1882 Pillsbury A-Mill, located on the east bank of the Mississippi River, operated two of the most powerful direct-drive water wheels yet built, each capable of generating 1,200 horsepower (895 kW). Upon Pillsbury's recommendation, the construction of another dam below the Falls added 10,000 horse power to the capacity already provided. A-Mill remained the world's largest flour mill for 40 years.

In Germany,Werner von Siemenshad invented an efficient dynamic-electric motor, with encouragement from his friend, master mechanic Johann Georg Halske. In 1879, they constructed a 328-yard-long (300-meter-long) narrow-gauge line for the park created near the Lehrte Station for the Berlin Industrial Exhibition. The train comprised a four-wheeled locomotive pulling three passenger cars capable of carrying six passengers each and operating on the third rail principle whereby the current was supplied by a third rail and returned via the running rails, the locomotive completing the circuit. No batteries were used.It was a resounding success. Werner von Siemens described the response to this attraction in a letter to his brother Carl Wilhelm:

Our electric railway [...] is quite a spectacle here. It is running even better than expected. In just a few hours' time, around a thousand people a day are being transported for a donation of 20 pfennigs to charity. The train carries 20 to 25 people and runs at roughly the pace of a horse-drawn tram. This is something we certainly can develop!

Based in England, Carl Wilhelm Siemens, known in Britain as Sir William Siemens, was interested in electrical agricultural machinery. In 1875, with the wealth accrued from Siemens Brother's long-distance telegraph sales, he was able to purchase "Sherwood", a 160-acre estate near Tunbridge Wells, Kent. No sooner arrived, Siemens erected a 6hp high-pressure Tangyesteam engine in one of the outhouses to power to an electric generator. By 1880, this was not only used to light "Sherwood" with arc lamps, and the grounds with arc lights, it also acted, through wires conducting it to various places, as the transmitter of power to perform work required on the estate, such as sawingwood, pumping water, chaff-cutting, and also electric light assisting the growth of plants, vegetables and fruits in a large greenhouse. In addition, the waste steam from the engine was condensed into a heater when the greenhouse took its circulating supply of hot water, thus saving the fuel that would otherwise have been required to heat the stoves.[13]

Sharing technology, Siemens' friend Baron Armstrong had also adapted his hydroelectric powerplant at "Cragside" for crushing corn, threshing, and ploughing and the nearby Cragend Farm used hydroelectricity for making silage.[14]

In 1881, a Siemens AC alternator driven by a watermill was used to power the world's first electric street lighting in the town of Godalming, Surrey. Sir William's experiments in electroculture were cut short by his untimely death that year due to an accident. In Germany, his brother Werner had taken out Kaiserlichen Patent 12869 for an electric cable plough, where the self-propelled plough was connected to an overhead power line.

The propulsion of the ploughs is done by secondary dynamos placed on them. The shoving forward of the plough by driving bars, moving forwards and backwards with shoes suitable to the soil, which grip the earth. The electrical connection of the primary and secondary by two stretched wire lines or cords either carried on contact rollers placed on the plough or wound on the same.[15]

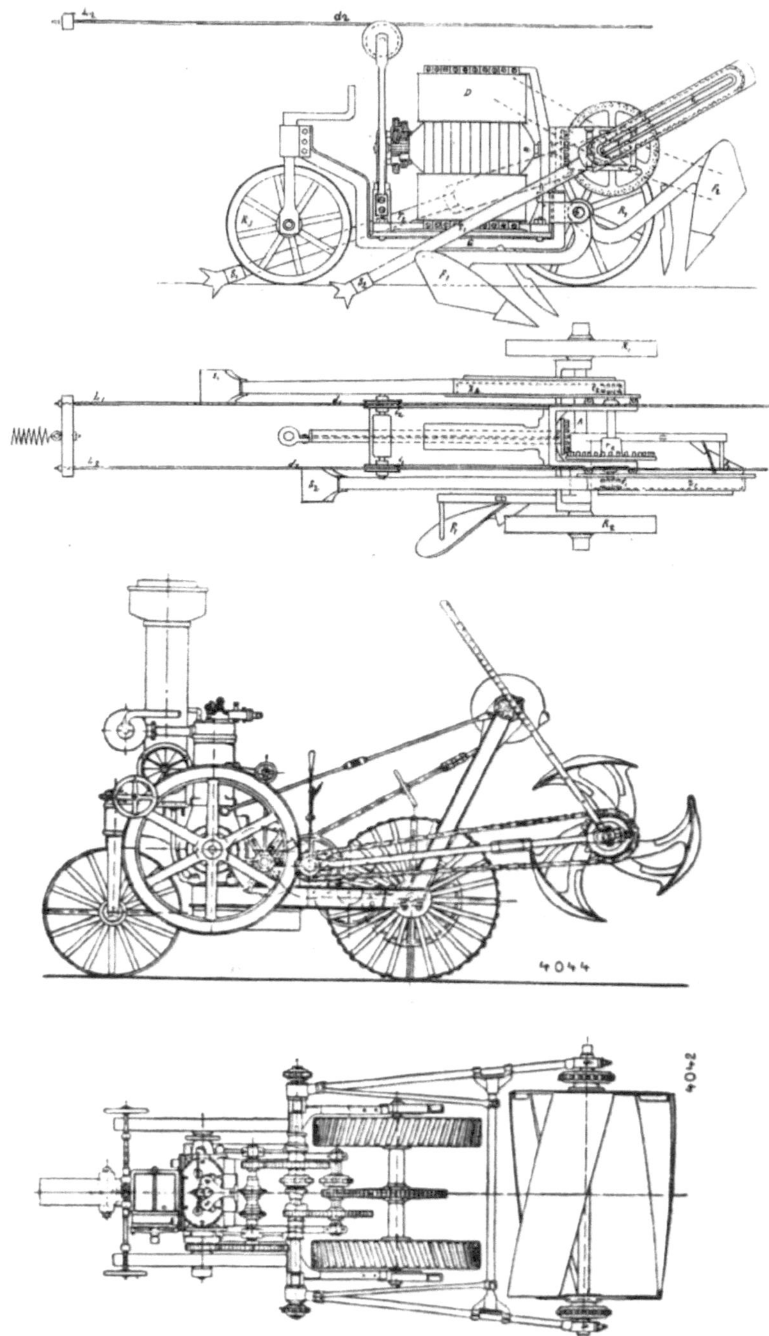

Werner von Siemens patented his design for an electric plough in 1880 (Siemens Archives).

In other parts of Britain electric power was also being developed. In 1880, William Traill and his brother Anthony built a generating station at the Walkmill Falls of the Giant's Causeway in County Antrim on the north coast of Northern Ireland. They installed 104 horsepower (78 kW) Alcott water turbines to produce up to 250 volts at 100 amps of electrical power for the The Giant's Causeway, Portrush, and Bush Valley Railway and Tramway Company. In 1888, Traill used the electric motor for threshing purposes enabling a large stack of oats to pass through a threshing machine of ordinary construction in about two hours. And in 1881,Robert Cecil, the 3rd Marquis of Salisbury, an amateur scientist with his residence at Hatfield House, in Hertfordshire, dammed the local River Lea to power a watermill which stood on the eastern edge of the park and used hydro-electric energy to light his house, to mill corn and to saw timber.

Werner von Siemens patented his design for an electric plough in 1880 (Siemens Archives).

Improving the Siemens design for a cable tractor would challenge German engineers during the next three decades. For example in the 1890s, Maschinenfabrik Borsig,a family concern in Berlin,tried out an electric motor unit and a movable anchor. In 1891, *The Rural New Yorker* reported on "an electric motor which could be located on a revolvable platform in the centre of a field to draw in ploughs on radial lines by means of cable." Trials carried out in 1895 by Messrs F. Zimmermann of Halle an der Saale, Saxony, Germany demonstrated how ploughing using electric traction cost one third per 2.5 acres hectare) than a traditional ox-pulled version. A Zimmermann was successfully trialled by Messrs Maguin and Bureau on the estate of M. Landrin at Bertaucourt-Epourdon in the Aisne region of France.

In 1882 the large Duchesne-Fournet bleaching ground near Lisieux France used this truck, powered by a Siemens electric motor and a stack of Force et Lumiere batteries to lay and take up cloth using a small winch. This was the world's first electric truck used for agriculture. (Musée EDF Electropolis, Mulhouse).

At the annual show of the Royal Agricultural Society of England, held in Plymouth from 21st to 27th of June, 1890 a few exhibits illustrated the increasing attention given by makers of engines and electrical machinery to the special requirements of country houses and large farms. According to the catalogue there should have been a self-regulating dynamo suitable for transmitting power to a motor for driving dairy and other machinery on a farm, as well as two electric motors suitable for driving farm machinery made by J.E Veale of St Austell. But only the dynamo was on view and not even working. Established steam machinery producers such as Fowler Ltd., Ransomes, Sims & Jeffries and Davy, Paxman & Co. were unimpressed.

Meanwhile in France, Camille Gouzi, proprietor of numerous small vineyards and farms in the commune of Millas (Eastern Pyrenees) having an area of altogether nearly 1,500 acres (607 hectares), was using a neighbouring stream, le Ruisseau de Millas for electric lighting purposes. In 1890, Gouzi applied electric power to the working of a grape-crushing plant. Besides providing the power for driving and lifting purposes, electricity powered pumps were used for irrigating the vines. In addition to this, threshing and other machines were powered by electro-motors. 180 light bulbs were distributed all over the farms and the length of the electrical wire connecting the buildings was no less than 62 miles (100 km).

In south-west France, on 22ndFebruary, 1895 electric ploughing experiments took place on the property of Félix Prat in Enguibaud, near Saint-Paul-Cap-de-Joux, in the Tarn. With electric power from an adapted watermill, Prat linked a Toulouse-built Pelous plough to a series of winches and almost 800 ft (240 m) of insulated overhead cables. With this extension it was possible to plough at a distance of 1476 ft (450 m) on each side of the power line, 2.5 ft (80cm) wide and 5,900 ft (1,800 m long) over 356 acres (144 ha). The speed of the ploughing, which dug a line 2 ft (60 cm) deep and 1.6 ft '50 cm) wide, was 85 ft (26 m) per minute. Thus 1 acre (4,000 m^2) was ploughed per ten-hour day requiring 131 kWh.

In his book "L'Electricitédans la Ferme", published in Paris in 1891 by the Librairie Agricole de la Maison Rustique, Maximilien Ringelmann devoted an entire chapter to La Ligne Electrique (Conducteurs, Les LignesAériennes).

In the French colony of Algeria, in Ben-Salah, close to Boufarick, a wine grower, Abel Pilon used a trolley plough on his vineyards with the innovation that the pulling cable was not attached to a fixed point, but to a mobile stem around a vertical axis controlled by a steering wheel which also controlled the depth of electric ploughing. The technology was reaching the a corner of the Empire;

With their factory in Ehrenfeld, near Köln, the company Helios Elektrizitaets AG., named after the Greek sun god, was one of the most innovative electric machinery companies of its kind around the dawn of the 20thcentury. Both electric trams and complete power plants were manufactured using the then-new AC system, which was put into operation across Europe. This included the provision of electricity on farmsat Quednau, Summern, and Crottorf in Germany.

At the 300-acre (121 hectare) Quednau dairy farm, in northern Konigsberg, there were two steam engines driving direct-current generators, one a 45 kW, 500-volt machine, and the other a 6.6 kW 220-volt machine, and in conjunction with these a storage battery. At night the engines were shut down, and the battery, which had been recharged during the day, was used to light the farm and furnish what power might be needed. For charging this battery, the 6.6 kW machine was driven to give 18-amperes at 320 volts. On this farm there were three motors: one, a 2hp unit drove a carrot-cutting machine and a straw-cutter;

another, a 15hp motor was used for threshing, driving a circular saw, pumping and milling. The third was also portable, but was usually kept in the granary, where it drove a mill. Alongside these, there was an electric trolley plough able to break up very hard soil, and cut 13 inches (33 cm) deep through it. Its speed could be regulated to 10 per cent of full speed by a resistance, and it passed from one furrow to the next almost automatically by means of two switches. At Quednau, which was comparatively a small enterprise, these motors had replaced the traditional twelve horses and eight men.

At Summern, a water-driven turbine was connected to a generator supplemented by a storage battery to help carry the heavier loads. On this farm, two motors were used, a three-quarter hp unit for driving a cream separator and a 10hp portable motor on a carriage that could be wheeled about to the different agricultural machines. These machines were grouped in such a way that several of them could be driven at once, ensuring economy in operating expenses and making the use of more than one motor unnecessary. The motor could be taken to one room then using a few metres of shafting be made to drive a carrot cutter, cutting 1430 pounds (650 kg) of carrots an hour, a machine for crushing cakes of linseed at the rate of 1100 pounds (500 kg) an hour, and a pump able to lift 353 cubic feet (1000 litres) of water an hour to the stables. In another place it was used to drive a threshing machine and at the same time a straw cutter, giving 330 pounds (150 kg) an hour.

The third station was at Crottorf in Saxony, covering one of the most fertile districts in Germany. Its position was nearly at the centre of a farming district 18 miles (30 km) in diameter, and in which there were twenty-seven towns. Electric power was obtained from a waterfall in the Bode River via three turbines geared to a common shaft to which was direct-connected a 500 kW three-phase, 7000-volt alternator. This shaft was arranged so that it could also be driven by a steam engine, coal being easily obtainable from neighbouring mines. Another steam engine, connected to a similar 500 kW generator, supplemented the plant; but the water-power was usually sufficient to supply the demand. The steam engines were able to carry the entire load when the water-power was not enough.

In each of the towns supplied by this plant there were transformer sub-stations and rotary converters which distributed the energy at low voltage. These sub-stations were supplemented by accumulators, charged during the day and assisting the converters at night. In the larger towns where both the lighting loads and the fluctuating power loads were heavy, separate circuits for lighting and power were used; but in small towns where there was only a light load, one circuit was used. One of the most interesting features at Crottorf was the portable transformers for taking energy from the high-potential circuits in the field at any spot where connections could be made, and delivering it at low voltage for power to drive agricultural machines[16].

Beside these, several experiments took place on the estates of enlightened, wealthy farmers in Germany, Austria, and Hungary.

In 1894, Ganz & Co. best known for their electric tramcars, equipped a farm at Jevišovice (German: *Jaispitz*) Moravia. A 30hp steam engine powered a dynamo from which two sets of cables went out about 3 miles (5km) in different directions, one leading to the sawmill and dairy, the other to a second dairy. There were three secondary stations with two 12hp and one with 10hp, the transmission being at 650 volts. Motor carts were arranged so as to be used as threshing machines at any point along the line and they drew their current by means of sliding contacts from the mains[17].Carl Heinrich von Teskow had a Borsig 500-volt direct current generator installed on his 250 acre (101 hectare) farm at Dahlwitz for lighting, pumps and trolley ploughing. Installed by F. Brutschke of Charlottenburg, it did not use cable drums, the redundant cables laid by hand in coils on two carriers fixed on the motor wagon. Three men were required to attend to the plough capable of a ploughing speed of 3 feet (1 metre) per second.

Cable machinery was not only in use in Germany. In Minnesota, USA one farmer used a cable electric power on his plough, harrow, his seeding machine, his reaper had to move a loaded wagon across a field in a straight line and back.

In June, 1895 F. M. Newton's Ltd. Electrical Engineers, Taunton, Somerset installed one of their turbines inside the newly built extension to the nearby Hestercombe Mill, which was renamed the Dynamo House and used as both a sawmill, to grind corn and to crush apples. Built by Gilbert Gilkes & Gordon Ltd. of Kendall, Cumbria, the innovative 6hp (4.41 kW) double Vortex turbine was the invention of Professor James Thomson (1822-92) of Queen's College, Belfast, who had patented his design in 1850.The *Vortex* had the great advantage that it could work on any head of water from 3 to 300 feet (1 to 90 metres), it was relatively small, and had an efficiency of 70 to 83%. If fitted with movable guide blades, it would maintain a reasonable efficiency with a low flow of water. The best steam engines of the day barely produced 10% and the newest gas engines only about 20%. Prior to Hestercombe, a Vortex had already been installedat "Cragside" in Northumberland, enabling the house to become the first in the world to be lit by hydro-electricity, using incandescent lamps provided by the inventor Joseph Swan. Another three were installed at the Duke of Devonshire's stately home of Chatsworth in Derbyshire, where water power from artificial Emperor Lake was used to power a Gilkes generator in an underground chamber, approximately 400 feet (122 metres) down the hill. Last but not least, in 1895 a Vortex was installed at Balmoral Castle, Aberdeenshire, Scotland, the favourite residence of Her Majesty Queen Victoria.

Under the direction of the Austro-Hungarian Imperial and Royal Agricultural Society, an international mart or exhibition of agricultural machines was held in Vienna from 4th to 7thMay, 1895. The prospectus included appliances used in all branches of industry concerned with agriculture such as brewery, forestry, milling and breeding trades. The exhibition also comprised electrical machines used in the service of these industries.[18]

For 1897, the German Agriculture Society arranged a competition power-driven ploughs and more especially, to encourage and develop the application of electrical power to the working of ploughs and other agricultural implements. Special attention would be given to the conveyance of the arrangements from field to field. At the trial, a piece of heavy and a piece of light land were assigned to each plough and the following points were noted: the time taken for the ploughing, the weight of earth moved, the consumption of fuel by the motor, the cost of ploughing including fuel, water[19].

Very well documented is the model farm developed by Count Vittorio Carlo Fernando De Asarta on his estate in at Fraforéano, in the Province of Frioul, north-eastern Italy.

De Asarta was born in Paris on 8th January, 1850 to Emanuele and Russian born Caterina Séraphine Eloy. He studied at the Imperial Lyceum and later at the École Polytecnique. After travelling around Europe, he devoted himself to agriculture, convinced of its economic and social importance. In 1870 he enrolled at the Higher Technical Institute of Milan, where in 1873 he obtained a degree in engineering. In 1882 he bought the run-down eighteenth-century Villa de Kechlerin the province of Udine. Its lands included about three thousand Friulian fields, covering 2470 acres (1,000 hectares) in the valley of the Tagliamento River.

During the 1890s Vittorio Carlo Fernando De Asarta pioneered cable ploughing on his estate in at Fraforéano, in the Province of Frioul, north-eastern Italy. (Collection Ciccinella Kechler).

Here with his young family, he began to rebuild the villa, its outhouses and prepared the land in preparation for cereal production (oats, rice and froment wheat). To plough the fields, far from an industrial supply of electricity, in 1890 de Asarta designed and built water-powered electric machinery. He took the current from a waterfall of the Barbariga Canal, a tributary to the nearby Tagliamento River, running the current along a conduit raised on wooden stilts. This then turned the curved blades of a French-designed Poncelet 25 ft (7m50cm) waterwheel which in turn belt-drove a German 1000-volt dynamo for a power output of 18hp carried by a lightweight four-wheeled cart. The electricity generated was used to drive a British built poised Howard plough over three acres (3 hectares) in ten hours per day. De Asarta also harnessed this energy to light his farm, to power the motors for the preparation of the fodder, for the watering of the manure, and for the mechanical maintenance workshop.

Asarta's German 1000-volt dynamo was hydro-electrically powered from the local Tagliamento River (Collection Ciccinella Kechler)

De Asarta thus achieved the first modern electric farm in Italy, the news of which soon spread across Italy and Europe. Despite this, he still encountered ongoing prejudice from traditional horse-loving farmers. But from respect earned, de Asarta was elected Mayor of Ronchis from 1889 to 1896, then Deputy for the college of Palmanova from 1897. The report of the parliamentary sessions shows his specific attention to the problems of the primary sector: from the need to promote more accurate agricultural statistics to the rules on sugar production; from health police issues to the problem of malaria. In 1902, he was designated a Cavaliere del Lavoro and elected Vice-President of the Società degli Agricoltori Italiani. Asarta died in 1909, most probably aware that gasoline-engine tow tractors were posing a threat to trolley-cable ploughs[20].

Among those who admired de Asarta was his fellow countryman, Emilio Guarini, born 1879, mostly remembered for his invention of the cordless repeater for Marconi's system of wireless telegraphy. But Guarini was also passionate about the potential of electric agriculture. In 1903, he published a seminal paper,

"Applicazioni scientifici dell'electricità nell' agricoltura" (Electricity as an Aid to Agriculture).[21] Guarini reported that when a plough was pulled by an electric motor the overall yield increased by 20%, that of beetroot by 26% and barley by 35%. He also describes two model electric farms where all machinery was belt-driven using steam-generated electricity from log-sawing to milk-churning, one dynamo driving 4 machines at a time; he also refers to working in Trani killing insects using electricity. Guarini concludes "Let us hope that our country with all its forces assists the direct initiatives to the application of electricity to agriculture."

During the 1890s, the idea of using cables for transport was popular because batteries had a low autonomy and were not reliable. A bicycle railroad was developed by William Boynton of McMorris Park, New York whereby a boat-shaped in-line two-wheeler deriving its power from an overhead electric cable had a potential speed of 100 mph (160 kph). In 1897, a prototype system was built to run from Bellport to the Sound across Long Island but was short-lived. Would-be aviators such as Sir Hiram Maxim in England were considering supplying their heavier-than-air flying machines with power from electric cables. Trolley boats were use on New York's Eerie Canal, the Teltow Canal in Germany and the Charleroi Canal in Belgium. Trolley trucks, based on the Schiemann system were also in use Hamburg and Leipzig, in Blankenese, in France (Lyon-Charbonnieres), Italy (Pirona), Norway (Drammen) and England (Leeds, Bradford). But most significant of all, by 1910 there were over 300 trolley tramway operators in cities around the world; so trolley-ploughs were merely another application.

In 1900, another application took place due to the extremely heavy fruit crop which occurred in Switzerland, prompting Scheller, a large boot manufacturing company based on Wohlen in the Canton of Aargau, to turn to account their day-time surplus of electric power. Using their techniques boot manufacture, they assembled a plant for the peeling, coring and slicing of apples, for the further treatment of these ring slices and for the baking of the prepared slices. The baking or drying of the slices was then carried out by placing the slices on grid-like trays which succeeded one another in the electric oven.

At the same time, Brutschke electric tractors were in use on farms near a sugar beet processing factory that had installed a generator to supply electricity for handling and processing the beet, but for much of the year there was surplus electricity available that could be supplied through overhead power lines to farms close to the factory. A report published in 1902 stated that several of the larger farms were using the factory's "off-peak" electricity to power the 220V motor on Brutschke tractors.

A Brutschke electric tractor of 1902"

Around 1908, Schuckert engineers Ernst Valentin and Franz Strarkloph designed a commercial vehicle powered by two rear-wheel hub motors. In addition to the normal steering wheel, there were additional left and right steering wheels on the outside of the truck, which could be operated by the driver beside the vehicle. The design was used as both a truck and a bus, its chassis manufactured by the Siemens-Schuckert plant on Motardstraße in Berlin-Siemensstadt and the motors by Siemens & Halske. From 1910, Siemens-Schuckert began to diversify and they looked into agriculture. Enter Konrad Victor von Meyenburg-Martin.

Von Meyenburg was born 1870 in Dresden, Germany, where his father Victor von Meyenburg worked as a mechanical engineer, but he was raised and educated in Switzerland. Konrad always stated that he had inherited his technical skills, as his great-grandfather constructed steam engines and railroad locomotives around 1800, without any formal education. Konrad's grandfather worked as a constructor with the Escher-Wyss company in Zürich, Switzerland. While still at school, Konrad fitted his bicycle with a small steam engine. Later he studied at the ETH (Electro-Technical High School) to become a machinery-constructing engineer. At the 1893 Chicago World Exhibition he impressed several American technicians and for some time he worked in Boston (USA), and later with Escher-Wyss in Zürich.

He then met up with Gottlieb König who was working on a "machine for ploughless tilling" which employed rotating blades to loosen the soil. Together, König and von Meyeburg developed the "König Landautomobil" rotary cultivator. However, von Meyenburg felt that there were several problems with König's swinging blades. During intensive studies he developed spring-mounted tines made from 0.2 inch (5 mm) thick steel wire. These flexible parts were able to move around obstructions such as stones or tree roots without being damaged which also resulted in a good crumbly soil structure. No one was able to tell Von Meyenburg what was the soil's optimal crumbling structure, so he talked to scientific and practical agricultural institutes in Switzerland, Germany, France, England and the USA, and also to the Russian institutes in St. Petersburg and Moscow, in an attempt to get better details.

In 1909, he patented the spring-mounted tines. He was able to raise interest for his ideas with young August Grunder, and together they undertook intensive tests in order to find the optimal shape for the tines. Simultaneously they started construction of a tiller tractor, which was patented in 1910 under the name "Motorwagen für landwirtschaftliche Arbeiten" (engine vehicle for agricultural labour). Both the tractor and the various forms of spring tines were made in an abandoned building in Zürich, Switzerland. In August, 1910 the tractor was shown to the home and foreign public at field demonstrations. This resulted in exhaustive descriptions in agricultural magazines. In 1911, the Patentverwertungsgesellschaft Motorkultur AG was formed, based at the Dornacherstrasse 160 in Basel. From 1912 to 1914, prototype models were taken on long demonstration tours into Germany, France, England and the USA. As a result, several companies obtained a manufacturing license. One of these was the Siemens-Schuckert-Werke GmbH, who bought the patent rights for Germany, Austria, Scandinavia and the Balkan countries. Siemens being an electrotechnical company, a 2 kW *electric* rotary cultivator was made. Indeed, 1,000 of these were manufactured.

Hydroelectric power for ploughing could come from watermills which had been in service for centuries. This was the case with the watermill at Katlenburg-Duhm in southern Lower Saxony which was first documented in the year 1326 and rebuilt in 1855. Fifty years later however, a 70hp Schuckert system was supplying 1000-volt tri-phase current via a 1300-foot (400-metre) drum-wound cable to neighbouring fields. A similar system was also used on the Hungarian estate of the late Count Lajos Batthyány, Prime Minister of the 1848 Government of Hungary before his assassination the following year. The Batthyány family used the tri-phase system in the localities of Steinamangar, Sárvár and Ikervar.

The Siemens-Schuckert cable pole could easily be replaced to neighbouring plots. The pole was 20 feet (6 m) high and the cable had a length of 82 feet (25 m), enabling tillage of a 13,000 $ft^{2(}$1200 $m^{2)}$ field from each pole position. A trailer behind the tiller was available to transport the pole and the power cables from one field to another. Where the machinery was not too far from the power supply, it could be directly connected to the electricity wires, which had 120, 210, 380 or 500 Volts AC, or 110, 220, 440 or 500 Volts DC. Siemens-Schuckert supplied special hooks, closing automatically to ensure a good contact. If the distance between the power supply and the field was so large that too much power was lost by the cable's resistance, high voltage was used, 5000 Volts AC or more. This solution required a transformer to bring down the voltage to the 1000 Volts AC which the 60 kW cable plough puller needed. The distance between transformer and puller was limited by the cable length of 980 ft (300 m). When furrows of 1640 ft (500 m) were ploughed, this meant that 74 acres (30 ha) could be ploughed from one connection point. The electric pullers usually had a petrol engine on board to provide propulsion from one field to another where no power connections were available between them.

In an anonymous work "Electricity on farms in Germany" published in 1912, the author reported that:

Siemens-Schuckert builds electric tractors that operate on the single-machine system and are designed in sizes up to 90 hp. Since ploughs are only operated in connection with overland centres, only three-phase motors are used. A complete plow set consists the motor car, plough cable, counter carriage, cable, rope carrier, field implement, cable car. The motor car stands on one side of the field to be ploughed, opposite it at distances up to about 400 m of the counter car. Between both runs the cable in which the field device is mounted. The motor car carries the drive machine, the electric motor, which transmits its power either on the wind drums, which move the cable, or after the application of a clutch drives the rear wheels of the car during self-drive. All engine parts and also the electric motor are protected by encapsulation. To operate the whole apparatus, two handy levers suffice. The control of the engine is done in the simplest way by a crank in the same manner as in the trams. Against incorrect operation and overstretching ensures an apparatus that automatically stops the engine when the pulling force on the cable exceeds the allowable level, e.g. when approaching larger stones, clogging the plough, etc. This reduces the risk of breakage if possible. The rope is made from the best plough steel wire. It is properly wound by means of cable guides on the drums, which are arranged exactly as in the steam ploughs. In addition, cable carriers keep the cable off the ground so that the wear of the plough cable is well within normal limits. The trailer is made differently depending on the nature of the field. The anchoring devices with which the carriage holds in the ground are in any case designed so that they reliably prevent the lateral displacement when the plough is pulled from the towing vehicle to the counter carriage. If the plough moves back to the motor car, the trailer car is relieved and is now moved as much as the work of the plough has progressed as a result of the rotation of the roller. The field implement is determined for each individual case on the basis of experience gained in this field, with particular emphasis on reducing power consumption. A voltmeter and an ammeter, which are attached to the towing vehicle, visible to the guide at all times, allow a permanent control of the operation. The power supply is a flexible cable, which is used in lengths of 300 to 1000 m. The insulation of the cable is made of the best, most durable rubber compound so that the operation can always be maintained even in the rain, etc. To protect against external damage, the cable is equipped with a suitable for the plough operation protection. It is thereby created a construction of the plough cable, which is characterized by the greatest resistance and excellent insulation by flexibility and low weight.

Between 1908 and 1914,1,600 Siemens-Schuckert electric ploughs were being used, particularly in the Prussian provinces of Posen and Pomerania. They could carry out deep ploughing as well as or better than steam ploughs and they promised to solve social and political problems such as the increasing labour shortages east of the Elbe. But they found little success because the fields had to be re-landscaped to make them work.[22]

As evidence of the believed potential of overhead electric cable farming, in 1899 the Meunier chocolate firm issued a series of 87 collectable artist cards in time for the 1900 World Exhibition in Paris. Entitled "France in the Year 2000", the cards feature interpretations by artists such as Jean-Marc Côté of the way people would be living. One of these, "A Very Busy Farm" shows a man controlling driverless trolley farm tractors like he might control the points on a railway system.

In 1899, the Meunier chocolate firm issued a series of 87 collectable artist cards in time for the 1900 World Exhibition in Paris. Entitled "France in the Year 2000", the cards feature interpretations by artists such as Jean-Marc Côté of the way people would be living. One of these, "A Very Busy Farm" (Musée EDF Electropolis, Mulhouse).

The first British-built electric plough was made by E.O.Walker & Co. Ltd., electrical, mechanical and ventilating engineers of Manchester for trials by a Mr Chorlton of Cotgrave Farm, some eight miles (13 kilometres) from Nottingham. Unsatisfied, Chorlton later converted it for timber haulage.

The Ontario Hydro-Electric Power Commission, created in 1906, provided access to electricity powered by Niagara Falls in Ontario, Canada, and even sent demonstration wagons to rural districts to illustrate the use of electric power in the household and its application to the driving of farm machinery.

1902: Englishman Mr J. Ellis Beatty with his mobile electric ploughing cable (The Museum of English Rural Life, University of Reading)

However, in the opening years of the 20th century, Robert Henry Fowler of Steam Plough Works, Leedstook ship to visit the United States where he became convinced that the internal combustion engine was the way of the future and would eventually replace steam or electric as the prime mover for agriculture and transport. In 1902, a first provisional patent was issued to Fowler's for an Otto Cycle or four-cycle gas engine. Later the same year, a three-cylinder engine based on this patent was built for the War Office for installation in an experimental motor car. In 1917, the Fowler Agri-tractor was introduced and over 200 were produced. Over in Germany, in 1903 even Helios of Köln who had been into electric trolley ploughing, attempted to launch petrol motor trucks from 10hp to 40hp, but ceased production three years later. From 1915, Siemens-Schuckert's von Meyenburg and Holdack also developed petrol-engined tractors with a range of 8 to 50hp which were trialled on the firm's testing grounds at Gieshof.

In the middle of the war in 1916, when the priority was how to move prototype caterpillar tanks and canons over the barren shell craters and trenches of northern Europe's war zones, the use of autonomous tractors still seemed a possible prospect, as proposed by J. Bligny:

Even if the electric ploughing would be at the same price as the ploughing with steam and even a little more expensive Would not the farmer have any interest in using the latter mode, since in the case of steam he is obliged to be concerned with the question of coal and water, while on the contrary he no longer need take care of anything for electric ploughing.[23]

Four years after founding the Ford Motor Company in 1903, Henry Ford finished his first experimental tractor on Woodward Avenue in Detroit, referring to it as the "Automobile Plow". Approximately 600 gasoline-powered tractors were in use on American farms in 1908.By 1925, Ford had built its 500,000th Fordson tractor taking 77% of the U.S. market. They were also built elsewhere in the U.S., in Ireland, England and Russia. In 1925 International Harvester produced the Farmall, the first high-wheeled, row-straddling tractor to cultivate row crops and cotton. The new tractor did its many jobs well and hence sold well, and by

1926, I.H. was ready for large-scale production at its new Farmall Works plant in Rock Island, Illinois. Although the Farmall never reached the per-year production numbers of the Fordson during the 1920s, it was the tractor that prevented the Fordson from completely owning the market on small, lightweight, mass-produced, affordable tractors for the small or medium family farm. Also in competition was John Deere with their *Model D* tractor; in France, the Renault *GP* and *PE*; in Italy the FIAT Trattori *703* and Ferguson in the UK. Between 1900 and 1960, gasoline was the predominant fuel, with kerosene and ethanol being common alternatives.

It was to be a century before the commercial electric agricultural tractor would be seriously re-considered, although research and development did continue.

Endnotes

[1]J.N. Nollet, "Recherches Sur Les Causes ParticulièresDesPhénomènesÉlectriques." Paris 1749.

[2]Pierre Bertholon, "De l'électricité des végétaux" Lyon, 1783.

[3]FernandBasty, "Nouveaux essaisd'electroculture." Bulletindela Sociétéd' étudesscientifiquesd'Angers, Nouvelle series 39:33-95. 1909.

[4]*Abraham Lincoln, "Address Given to the Members of the Wisconsin State Agricultural Society and Citizens of Wisconsin," 30*[th] September, 1859.

[5]L.T.C.Rolt, "Great Engineers", 1962, G. Bell and Sons Ltd,

[6]As Fowlers were also a major exporter of engines to the former British colonies in the 1800 & early 20th century a number survive in places such as Australia and New Zealand in preservation and in India, Pakistan, Africa and South America as rusting remains. Some of these are being repatriated to the UK and restored by collectors as few original UK engines remain to be restored.

[7]A. Dubois, "La moisson à la lumière électrique", *Journal d'agriculturepratique*, 1878, vol. 2, p. 188-190.

[8] Article by Péronne"On Nous écrit" Le Messager de la Marne, 23rd May, 1879.

[9]Gustave Cabanellas, "Electric transport of remote mechanical work, ploughing applications, Chrétien and Félix system" La Lumière Electrique, 15th June, 1879; and "Early Uses of Electricity in American Agriculture", Clark C. Spence, Technology and Culture, Vol. 3, No. 2. (Spring, 1962), p. 143

[10]A. Niaudet, "Le LabourageElectrique" La Nature 1879 N°314-339; Jean-Paul Bourdon "Des charruesbranches" La France Agricole, 28th March, 2011.

[11]Stand 457 also exhibited a railway with electric motor, equipment for working in the mines and the exploitation of forests.

[12]Pierre Richard, "150 Ans de Sucre à Sermaize" (1854-2004) pp78-80.

[13]C. William Siemens, "On Some Applications of Electric Energy to Horticulture and Agriculture," The Scientific Works of C. William Siemens, Volume II London, John Murray, 1881.

[14]"The Thrum: A Guide to Rothbury and Surrounding District "by an unknown author, dated 1885

[15]*The Scientific and Technical Papers of Werner Von Siemens* John Murray London, 1895 (Volume II pp. 456-457).

[16]*The Electrical Age*, Volume XXXIII, July – December, 1904.

[17]"Electric Supply for Agricultural Purposes" The Electrician, 14th December, 1894.

[18]"Electricity in Agriculture," The Electrical Review, 1st February, 1895.

[19]*ElectrotechnischeZeitschrift* 12th November, 1896.

[20] E. Sagnier, "Il Tenimentodi Fracoreano ; Excursionid' unfrancese in Italia" Giornale di Udine, 9th September, 1903; "Il Tenimento de Fraforeano" LAmico del Contadino: periodico agrarian della Domenica, Udine, 8th September, 1903.

[21] *Emporium* Volume XVIII N°104 pp 121-140; republished in Nature, 19th March, 1904.

[22] Augustus Petri, "The Introduction of the Electric Tiller", Electrical Engineering Journal, 19th March, 1925.

[23] J. Bligny, "Des applications del' électricité, notamment à l'agriculture et au labourageélectriques des coopératives et régiesmunicipals" La Technique sanitaire et municipale, 11th year, n° 5, May 1916, p. 121.

2

Towards the All-Electric Farm (1920-2000)

In 1920, although electric vehicles were not used on the American farms about 5,000 electric industrial trucks were produced by 12 American manufacturers, while estimates of the total fleet ranged from 6,000 to 15,000, most of them built during the war. Most used an exchange charging system. Among the manufacturers seen in and around cities and towns were the Commercial Truck Company, the Lansden Company, the Walker Vehicle Company (now owned by Commonwealth Edison), the Ward Motor Vehicle Company, the Steinmetz Electric Motor Corporation, Walter Motor Truck Company, and the Baker R L Corporation.One step away from agriculture, Charles A. Ward of New York City, was head of his family's baking company, made up of the Mackey Company, the Ward Corby Company, the Ward Bread Companyand the Ohio Baking Company of Cleveland. In 1911, Ward had decided to replace his horse-drawn delivery vehicles with electric trucks, thus doing away with the unsanitary stable hitherto part of or adjacent to bakery.[1]

Alongside this, the challenge of using stationary electric agricultural machines continued to be met. Two innovations which entered into dairy farming were the milk cream electric centrifugal separator and the surge milker. The Babson brothers (Henry, Fred and Gus) of Chicago were friends of and dealers for electrical battery inventor and entrepreneur Thomas Edison, and importers of Belgian-made Méllote separators. In 1916, the Babsonsproduced their own Pine Tree milking machine. While the overall design was similar to others, the electrically powered Pine Tree Pulsator outperformed any other pulsator made. These machines were powered by a 32-volt Delco system with 12 batteries near the barn. Pine Tree Pulsators were simple enough that the farmer could maintain them himself. The Pine Tree Pulsator was patented in 1921. The problem with this unit was that it was difficult to keep clean.

In the Fall of 1922, Herbert McCornack invented the Surge Bucket Milker, which used a pulsating tug and pull movement like that of a calf. The new milker surpassed all expectations. The biggest advantage was the ability to easily clean and sanitize the Surge. Though as with older systemsthe bucket would be set on the ground, with the Surge there were only 4 inches (10 cm) from the teat to the pail, unlike the long milk tubing of the earlier standing models which was highly prone to contamination. Babson introduced the Surge milker in 1923 and it met with success across the USA. It could be run using a petrol engine, but with an electric motor the energy bill was halved.

By the late 1920s, most of the electric ploughing machines around the world had come from the Siemens-Schuckertwerke auto plant and dynamo plant factory, although two foreign factories built versions according to the same patents and licenses. During this period the firm considered whether it should make its machine as versatile as possible or modified to specific work: either transform it into a caterpillar tractor to pull heavy loads or into a machine with a ploughing tail. They achieved this using gears and screw spindle, from worm drive directly attached to this crank arm, which caused the lifting and lowering by means of a cable. The SS combination was particularly useful for forestry work. They received large orders from the Soviet Unionfor the cultivation of steppes and difficult marshlands for processing rice and similar crops.

1920 Rotary tiller, with block motor Grunder system, 2PS (Siemens Archives).

In France, in October, 1921 the Agricultural Office of France's South-West Region organised electric trials at the Ondes School of Agriculture, 25 km (15 mi) from Toulouse and 3 km (1.8 mi) from Castelhaud' Estretefonds. A large transformer received its hydro-electric power from the Orlu River, linked by winches to poles along the fields and vineyards to power the ploughs. Albert Douilhet of Bordeaux, the inventor of the Aldo harvester, designed electric tillage winches that "cost, consume and wear less than a direct tractor" and "pull everything, everywhere." In 1927, trials of electrically-driven and other ploughs took place at Aubergenville, in the Seine-et-Oise Department, under the auspices of the French Minister of Agriculture. With electricity provided by Nord-Lumière Triphasé, three sizes of electric tractors were tested: 80hp, 150 hp and 35 hp machines made by either J. Estrade of Carcassone or the Société Géneraled'Agriculture (SGA), both using the traditional Fowler steam cable system. Between the two wars, 6,000 hectares(15,000 acres) were electrically ploughed in the Aisne and Oise region.

Electric ploughing limited the farm to the electricity brought along by a chain of poles. (Musée EDF Electropolis, Mulhouse).

The front cover of the French magazine "Science et Vie" published in September 1927 showed an electric tractor system (Musée EDF Electropolis, Mulhouse).

25 tonne 250 hp cable tractor as built by the Société Genérale Agricole (Musée EDF Electropolis, Mulhouse).

In 1926, Franz Joseph Gillain Remy of Fosses-les-Namur, Belgium concerned about wear and tear, obtained a patent for trolley ploughing, where the cable consisted of a band of insulating materialof flat cross section having parallel edges formed with recesses, electric conductors being mounted in the said recesses.(US. 103,132).

In 1927, two Scottish cousins of the Fraser-Mackenzie lineage, 'Dick' and 'Jack' who had emigrated to Southern Rhodesia, (modern Zimbabwe) became the first to use a steam-generated electric plough shipped out from Great Britain to start farming in that area.

One of the pioneers of electric agricultural machinery for forestry work was Andreas Stihl. Born in Zurich in 1896, he was educated in Germany. From 1915 to 1917 he served in the Kaiser's army and was injured several times. Stihl then studied mechanical engineering at the Technical College in Eisenach, where he passed the engineering exam in 1920. From 1923 to 1926, together with Carl Hohl, he ran a small engineering office which was active in the field of steam boiler pre-firing systems without great success. In 1926 Stihl founded A. Stihl Maschinenfabrik in Stuttgart to make washing machines and the factory also built pre-combustion systems The innovative Stihl also tackled the question of how to simplify and accelerate heavy forest work and he invented a "Cut-off Chainsaw for Electric Power" ("Die Kleine Stihl Elektro"). This saw weighed in at a hefty 64 kg (140 lb.) and had a 2.5 cm (one-inch) gauge chain with handles at either end. Due to its bulk, it required two people to operate. Some versions had a wheeled generator which had a power cord to run to the electric saw.So in 1929 at the Leipzig Trade Fair, Stihl abandoned electric power andlaunched the world's first petrol-powered chainsaw, named the 'Tree-Felling Machine' and abandoned electric power. Within nine decades, the small mechanical engineering workshop developed into a global market leader in petrol-power saws and tools with more than 16,000 employees.

The various systems of electric ploughing which had been devised so far could be divided in two classes: first, funicular systems which transmitted motion to the plough or ploughs by means of steel ropes moved by winding drums or the like, such drums being actuated by electric motors and remaining on the borders of the ground to be ploughed, and second, direct traction systems in which the plough or ploughs were directly pulled by an electric tractor driven by a motor on board, where the tractor ran to and fro on the ground acting like a common steam or petrol tractor. In both cases the electric energy had to be transported to the motors by means of a cable or a system of overhead contact lines. If the cable was used, it must wind and unwind on a revolving drum in order to follow the movements of the motor, lying on the ground between the drum and the motor. If the motor was placed on a tractor the cable would drag on the ground and be subjected to intense friction and wear.

In 1927, a different approach was taken by two brothers, Silvio and Mario Mazza and Luigi Bolledi of Carpenato, a commune in the Province of Piacenza in the Italian region of Emilia-Romagna. The fundamental principle of their new system consisted in the fact that the cables for transmitting the electric current to the motor which was mounted on the tractor, instead of being rolled and unrolled on a drum and dragging on the ground, were kept suspended in the air by a balloon filled with hydrogen or other lighter than air gas. The electricity generated from a motor car was supplied via a pole which could rotate depending on the position the plough. The Mazza-Bolledi system was published both in Italy and in France, an artist's impression appearing on the front cover of the N°133 July 1928 edition of *Science et Vie*.

The front cover of the French magazine "Science et Vie" published in July 1928 showed the Italian innovation of suspending the cable of an electric plough using a balloon. (Musée EDF Electropolis, Mulhouse).

Perhaps the most passionate advocate of electric agricultural machinery during the 1920s was Richard Borlase Matthews of Felcourt Farm, which he renamed "The All-Electric Farm". Borlase was born in Swansea, Wales in 1878, one year before the world's first electric ploughing experiment had taken place in Sermaize les Bains, France. He went to sea at the age of 12 with a shipping company in which his father had a financial interest. He then served an apprenticeship in a South Wales tin plate works after which he took a degree in electrical engineering at London University. In 1904, Borlase travelled to the USA where he had secured a departmental appointment with the General Electric Co., at Schenectady, New York, later becoming an electrical engineer in Amsterdam for the Edison Electric Light and Power Co. and then returned to England working for London Underground Railway. In 1908, Borlase obtained a patent for "Improvements in or relating to Means for Applying Motive Power to the Driving of Machinery."

In 1912, Borlase, who had witnessed the Wright Brothers' first hour-long flight, published "The Aviation Pocket Book" printed by Crosby Lockwood and Son) containing 'the theory and design of the aeroplane, structural material, examples of actual machines, meteorological data, military information and signalling'.

This book was reprinted no less than five times between 1914 and 1920.At least one bulk order for this book was received from the American Air Force. During World War I, Borlase was attached to the Air Ministry in Britain where he worked on aircraft design. After the War, heco-authored" The Aircraft Identification Book", a 'concise guide to the recognition of different types and makes of all kinds of aeroplanes and airships'.

Following the Armistice, the Ministry of Agriculture appointed an Electro-Culture Committee to investigate the theories of electro-culture. It was found that where electrical current was used there was generally a substantial increase in yield, but interest in the experimental work soon waned because although the electrical effect was real, economically the cost of installing the system on a large scale was too high, so the project was abandoned. However, to establish, expand and evaluate his own theories on the use of electricity and farming, Borlase purchased several lots of the 312 hectare (772 acre)Felcourt Estate in East Grinstead, situated between East Grinstead in Sussex and Lingfield in Surrey. In this he was financed by his wife Esther née Barratt from a wealthy long-established family of London silversmiths. Borlase was convinced that the use of electricity was the way forward for farming. During the next decade, assisted by William Henry Pike and his son Harry, he developed and implemented 67 different uses for electricity in farming practice. It was in and around the farm buildings that electricity had its widest uses. Borlase developed the Witton 5 hp portable farm "Drumotor". This could be moved around the farm, plugged into a suitable socket and used to drive all sorts of machines via a belt. Where machinery was in constant use such as in the milking parlour and dairy, permanent electric motors were installed. Even a simple task such as inspecting livestock at night was so much easier with electric lighting. The farmer could enter the building at one end switching on the lights and exit at the other end, turning the lights off as he left.

During the 1920s and 1930s Richard Borlase Matthews campaigned tirelessly for the adoption of electric agriculture machinery. (Felbridge History Group).

From the outset it was run as a commercial enterprise. Staff were paid according to the hours worked which could be accurately measured as they had to clock in.Some of the resulting developments were marketed by the British General Electric Company, but some were inevitably dropped from farming practice. His vision of farming was too far ahead of his time.

The electrical supply for Greater Felcourt was generated by utilising water power from Wiremill Lake to drive a turbine. The current was carried some distance to the farm by overhead cables and although the initial cost of installation was high, the power supplied was 'satisfactory' and was practically the last cost made on supplying electric power. It created a 400 volt, three-phase current which could be supplemented by an engine driven dynamo on the farm. Some years later when a mains electricity supply became available in the early 1930s, Borlase still chose to use his initial electricity supply in preference to the mains supply which was only used at night when it was at its cheapest.

By 1921 Borlase had already given some considerable thought to the uses of electricity on farms, producing a table of uses broken into four sections: In the farm buildings: electric light in cow barns, or byres, for intensive feeding of sheep, pigs, poultry and other stock during winter and general lighting of buildings and yards. Food preparation for livestock: electric driving of chaff cutters, chaff dust extractors, root pulpers, slicers and cutter, cattle cake breakers, corn crushers, kibblers, rollers, grinders and grist mills, disintegrators, corn or maize shellers, huskers and shredders, electric driving of food sifting and mixing machine, meat grinders and mincers. For Dairy: Electric driving of milk or cream separators, cream ripeners, butter churns, butter workers, butter cutting and printing machines, butter tampers, refrigerators, ice breakers, ice cream making machine, milk circulators and pumps, cheese curd breakers and curd grinders, casein grinders, milk churn transporters, milk churn elevators, milk bottle cleaners and fillers, milk and cream pasteurisers, milk clarifiers or centrifuges, centrifugal fat testers, milk cooling and circulation pumps. Electric sterilisation of milk by electrolytic bath and mercury vapour lamp. Electric heating of incubators for testing bacterial content of milk etc. For Poultry: electric light in laying houses (to increase egg production), electrically heated incubators, hovers and foster mothers, high tension electrical treatment, electric fans for circulating air in incubators, testing eggs, bone grinders, grain crushers, grinders, kibblers and grist mills, seed cleaners and sorters and mixing machine. For general electric power: Electric pumping of water for domestic use and for stock (open tank and pressure system), pumping of liquid manure, pumping of sewage, white-washing machines, portable electric winch, hay hoists, hay elevators and transporters, food transporters, cart loaders, horse clippers and groomers, horse tooth grinder, sheep shearers, milking machines, grindstone (also mower knife grinder). There was an electrically-operated workshop for repairs to farm machinery, including grindstone, emery wheel, drilling machine, portable electric drilling machine, lathe, forge, blower electric soldering iron, electric glue pot, etc. Also, an electric thermo-couple long distance thermometer for tuberculosis test of cows and pigs, branders for marking cattle, last but not least electric clocks, bells and signals.

On the farmland electricity was used for irrigation and manure distribution; ploughing, cultivation and harrowing was by electric tractor or electric haulage set; hay mowing with electrically operated mower, hay making with electrically operated machines, hay drying with electrically operated blower; corn cutting and binding with electrically operated harvesting machine, in addition to which there would be electrically operated potato digger, fence post driver, stump and root pullers. As for the treatment of growing crops, electroculture, Borlase proposed high tension and high frequency electrical discharge; ozone treatment, mercury vapour lamp treatment: for better growth, to destroy insect pests, to strengthen the plants and so enable them to better resist the vagaries of the weather – as well as the wet and dry spraying of hop vines, fruit trees, etc. For transportation: electric tractors and electric milk vans and electric lorries for general transport. When it came to crop treatment, seeds could be prepared by an electrolytic seed bath (to produce heavier crops), treatment of seeds with ozone or mercury vapour lamp, the electrical germination of seeds, electrical seed dressing, heat treatment. For gathered crops: hay drying by aid of electrically driven fan, desiccation of vegetables and fruits, pulping of fruits for supply of jam makers, silage, automatic silage stackers, cutting and delivery by blowing up elevator, threshing machine feeders, trussing straw, bailing straw, chafing straw and hay, grain grading, potato sorting, winnower or tanner for crushing and grinding cereals, hop drying shredders and dryers (centrifugal), alfalfa mills, flailing machines, malting, clover pea and bean pullers, killing mould growths, seed cleaning and dressing, grain elevators and sack elevators. For rural industries, Borlase listed peat and timber, drying and seasoning, sawing and working, cider milling and pressing, wool spinning and pressing, weaving and knitting, clog making, toy making, hop drying, drying of spent hops for litter, drying of spent cider apples refuse for cattle food, drying of spent sugar beet refuse, potatoes etc., concrete mixers, stone crushing, preparation of nitrogenous manures by electrical discharge, brewing beer and he manufacture of alcohol. Finally, in the farm house itself, electric lighting, heating, cooking and domestic electric power.

In short, Borlase believed that any farm machine that revolved or had revolving parts could be operated by an electric motor, more effectively and more economically than by any other means.

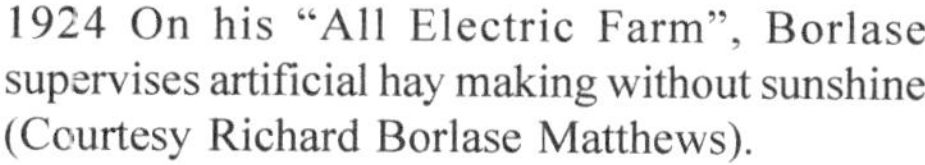

1924 On his "All Electric Farm", Borlase supervises artificial hay making without sunshine (Courtesy Richard Borlase Matthews).

1924 Borlase patented DRUMMOTOR powering a chaff cutter and electrically powered feed grinder (Courtesy Richard Borlase Matthews).

Greater Felcourt was only the second farm in Britain to trial an electric tractor. Ploughing was carried out by using a two-furrow balance plough pulled by cables attached to a portable electric motor.It was claimed that this was better for the soil as there was less compaction but the increased efficiency of tractors meant that this idea was not adopted. In papers delivered by Borlase at Nijmegen, Holland, in 1921 and Cardiff in 1922, he outlined several methods of electrical ploughing but by 1925 concluded that if pursuing ploughing by the use of electricity a small electric plough must be controllable by one man, advocating the use of a tractor type machine with an attached cable drum, with large pneumatic tyres on the front wheels and creeper tracks on the back to prevent pressure on the surface of the ground, instead of poles supporting cables. A tractor of this type, he concluded, would plough an area of 40 to 60 acres (16 to 24 hectares) from a single contact in the middle of the field.

Running hand in hand with his experimental use of electricity in farming practices, Borlase also became a consultant electrical engineer, delivering several lectures on the subject, his message reaching all over the world with newspaper articles appearing as far away as Australia and New Zealand. Borlase also wrote several books on the use of electricity, particularly in farming practices, which included "Electricity for Everybody: A Handbook for Central Station Engineers and All Users of Electricity", already on its third reprint in 1924and "Electricity, Civilisation and the Countryside" published in 1927. Borlase also contributed articles to such publications as *The Journal of the Royal Agricultural Society of England* and *The Electrical Review*. In 1926, at the height of his experimentation of farming with the use of electricity he wrote a book called "Electro-Farming, or the Applications of Electricity to Agriculture" which was the most comprehensive report on the subject, compiled from information he had gathered from extensive trips abroad and from his own farming practices at Greater Felcourt.[2]

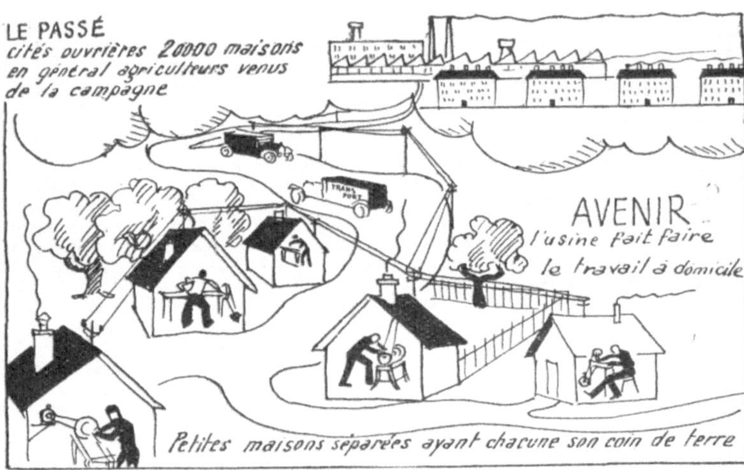

By the 1930s the rural electrification programme was in full swing. This cartoon was published by W. Bordigoni in "Le Génie Rural" magazine in November 1934. It was captioned "A new world where citizens, instead of living close to the factory in the city, will be part craftsmen working from home, and part farmers each one living from their little parcel of land and linked to the factory by electricity. And as for motor cars, isn't there here an economical hygienic and social ideal that one can envisage from now on?"

The Rothamsted Experimental Station is one of the oldest agricultural research institutions in the world, having been founded in 1843. It is located at Harpenden in the English county of Hertfordshire. On 9th February, 1928 a conference was held at Rothamsted, entitled "Power for Cultivation and Haulage on the Farm", including discussion of the electric trolley plough and a system in Italy for attaching the cable to a balloon, thus raising it out of the way. In 1930 Borlase prepared a scheme for performing all the operations about the Rothamsted farm buildings by electrical power. The matter went into abeyance until early 1932 when Sir Huge Hirst generously provided the sum of £500 for equipment and arranged the services of F.E.Rowland, agricultural expert of the General Electric Company. Thus the farm buildings were electrified and work continued over a period of three years (1933-1935) while Rothamsted made comparative evaluations of a 20hp electric motor, a new tractor and an old tractor.[3]

In 1920, the McDowall family had bought Mungoswells, an arable farm north of Haddington in the 'bread basket' county of East Lothian, Scotland. In the late 1920s and early 1930s Major Andrew McDowall developed a 12.5 hp electric tractor, which was built in the workshop at Mungoswells. The tractor was based on a three-wheeled chassis, with ploughs attached to the front and rear working alternately as the tractor travelled backwards and forwards. Power was supplied through a cable plugged into a power point that could be repositioned as the ploughing progressed, allowing the tractor to travel about 1,300 ft (400m) in each direction. Major McDowall claimed that ploughing with electric power was half the cost of using an ordinary tractor, but his request for government finance to enable him to carry out further research met a polite, but firm, refusal.

In October, 1928 a pioneer demonstration of electric threshing was given at The White House farm, Pettistree, East Suffolk, owned by Mrs F. Pearson. The machinery was supplied by Messrs Ransomes, Sims and Jeffries Ltd. of Ipswich, using electricity supplied from the Wickham Market substation of the Distribution Company of Leiston. The driving unit was a 25hp three-phase 400V drip-proof Ransomes motor complete with waterproof control gear, mounted on a four-wheel undercarriage with shafts for transport. This belt drove a Ransomes *A 11-54* threshing drum and chaff-bagger, fitted with ball-bearings throughout. A group of local and regional farmers was impressed to see 1400-1500 kg (1.4-1.5 tons) of barley were dealt with per hour, far faster than thetraditional cumbersome steam engines.

In November, 1934 *La Genie Rural* published a special issue entitled "L'Electricité Rurale – Electricité à la Campagne" in which it pointed out the many ways in which farmers could use electricity: forroot cutters, straw choppers, crushers, mills, oilcake and fertilizer grinders, sorters, pumps, threshing machines, creamers, churners, kneaders, and log saws. It also reported on short documentaries, silent and sound, made by the Societé pour le Development des Applications de l'Electricité (The Society for the Development of the applications of Electricity) APEL, encouraging people to use electricity on farms throughout France.

Italian Fascism made wide use of hydro-electrical power in the 1930s in an attempt to wean the country off imported fuels, when large quantities of cheap power were obtained from alpine dams. Since 1907, Professor Camillo Sacerdoti's Milan-based company had been active in various fields of industry such as in the iron and steel industry, rolling, lifting and mechanical processing.Camillo now studied electric power in ploughing and in 1932, finalised the *Elettra*, a 3,500kg (7,700 lb) tractor powered by a 25 hp electric motor, which was hitched up to a long supply cable on a self-winding reel. The idea got no further than the experimental phase, especially because Italian cables at that time were unreliable. At about the same time, AMA, the Anonima Macchine Agricole company, also based in Milan, designed a very similar tractor using a 17 hpelectric motor with 220 V cable power supply from a 1.40 m (4.6 ft) diameter motorised drum, equipped with a clutch for unrolling and of a motor (0.5 hp electric) for rewinding. It too remained a prototype. Italy never did quite achieve that self-sufficiency.

Over in the USA, by 1930, only 10% of the rural population had electricity. To rectify this, in the mid-thirties the Rural Electric Cooperative Associations (REA) brought electricity to farms although all buildings had to be re-wired for a stronger voltage. Irrigators began using large electric motors. Magazines such as the Nebraska-based Electric Farmer and Electricity on the Farm published articles with hints about using the new power for farm work and housework. When electricity arrived, the hay bale spear, pulley and track system were replaced by long motorized bale conveyors known as hay elevators. Companies making the electric hay baler included New Holland, Cardinal, and John Deere (*Model 33*). Belt-driven, they used a 1.5 hp 115 V electric motor and a 0.5 hp Marathon 1 Phase 115/230 V motor, often supplied by Delco (Dayton Engineering Laboratories Co.) When farmers met problems with the engine/generator, they reverted to the kerosene they were using in their tractors.

In 1930, Milton H. Arndt, sales manager in the brooding division of Kerr Chickeries in New Jersey had an idea for increasing egg output in the form of vertical hen houses. At his experimental New Era "Demonstration" Breeding Farm in Trenton, New Jersey, several large electrically heated and lit hangars, battery cages were fitted with a sloped floor that allowed eggs to roll to the front of the cage, where they were easily collected by the farmer and out of the hens' reach. Electric conveyor belts under the cages removed manure, so providing better air control quality and reduced fly breeding. From 1,000 hens, Arndt progressed to between 20-25,000 chickens. He discovered that playing music on the radio helped increase output. Between 1931 and 1933, Arndt published three books on the subject of 'Battery Brooding'. He also published articles, for example on that appeared in the January, 1937 issue of *Modern Mechanix* entitled:'BE INDEPENDENT! Own and Operate an "Indoor Poultry Farm"'. Arndtsold his chicken raising equipment locally and nationally through mail order catalogues. One million chicks were dispatched to poultry farmers each year. By 1940, there were nearly 40 million hens laying eggs in brooding cages across the USA. Condemned as cruel, the system has since been replaced by free-range farms, although not everywhere.

From 1934, the Electric Home and Farm Authority (EHFA) of America worked with manufacturers, utilities, and retailers alike to offer electric appliances to low- and moderate-income families through inexpensive purchase prices and affordable financing. Customers needed only a small down payment and had up to five years to pay off their appliances. Through the EHFA, the government transformed the domestic appliance business, and the home kitchen during the Great Depression by opening what had been a luxury industry to a mass market in the South.

An anonymous poem published at the time sums it up:

> Electricity is a servant, make it work for you. Then baking days won't be so hot, or washdays be so blue. Your cows will be contented, with a milker fine and bright. The kids will like the music, from the radio at night. Your feed will be ground easily, your baby chicks kept warm. The whole family will be happy, with electricity on the farm.

In the USSR, the large hydroelectric dam on the Dnieper River in the Dnepropetrovsk province made it possible for Soviet farmers to use electric farm equipment such as the SS two-way plough, about which an article was published in *Science News Letter* 23rd October, 1937.

Trials were continued in England at the Rothamsted Experiment Station by Messrs Cashen and Keen. The experiments were on threshing wheat, oats and barley, and on the grinding of barley for meal. Motors used were a 20 hp General Electric portable unit and a 20hp Harvester. They also used a new International Harvester Company tractor and an old International in use at Rothamsted since April 1928 and still in fair condition after 7,000 hours' work. Sixteen two-hour runs were made, of which seven were with the motor, six with the new tractor and three with the older unit. Special arrangements were made to obtain accurate measurements of the electricity, and the fuel and lubricating oil used in each experiment, and in addition records of the following five areas were taken: the weight of produce as first and second-grade grain, offal, chaff, cavings and straw; time and labour required to line-up motor or tractor with the threshing machine; the cause and duration of any stoppage; revolution speed of driving pulley and threshing drum; petrol required for starting and warming up the tractor before turning over to paraffin. The Marshall threshing machine used had a drum width of 122 cm (48 in) and the optimum speed of the drum was given at 1,240 rpm:

> The percentage of second grade grain and offal was slightly higher with the motor, and although a constant feeding rate to the thresher was aimed at, the rate was some 4 percent loss with the motor due, it may be suggested, to the psychological effect of the quieter running. But these differences are not significant and may well be due to variations in factors such as stack height, sheaf weight, and the fact that any one crop provided only sufficient material for four to five experiments.[4]

However, support for 'electro-farming' had waned by the early 1930s, primarily due to the cost and feasibility of installing an electrical supply to farms, many of which were situated in isolated locations and preferred to use horses and human power. At the same time, further active developmental work on Borlase Matthews' farm at Greater Felcourt ceased and he leased it to tenants on the condition that they continued to use his established electrical apparatus and methods of farming. It became known as the GEC Commission Electrical Farm.

From 5th to 9th July, 1938 The Royal Agricultural Show was held in Cardiff, Wales. A crowd-pulling feature alongside the Royal Enclosure and Flower Tents was called "The All-Electric Farm", perhaps inspired by Borlase' prototype. It covered some 2,230 m^2 (24,000ft^2). The initiative was that of the South Wales and Monmouthshire Area Committee of the British Electrical Development Association, directed by Lawrence Howles. A scale model of the All Electric Farm had been exhibited earlier that year at the Welsh Industries Fair.

22 TUESDAY, JULY 5, 1938 WESTERN MAIL AND SOUTH WALES NEWS

THE ALL-ELECTRIC FARM

Wonderful Exhibit OF THE Electrical Development Association

See the THOMPSON BAYLISS BUILDINGS in the ALL-ELECTRIC FARM

Dormerwood Bungalow

£165 ERECTED (on client's own foundations)

ACE POULTRY HOUSES

DRAYTON PIG HUT.

KENNELS 37/10.

GREENHOUSES

AVIARIES. GARDEN SHELTERS. GARAGES. STABLES. KENNELS AND FARM BUILDINGS OF EVERY KIND SUPPLIED TO YOUR REQUIREMENTS. 300 MILES FREE DELIVERY.

Catalogues Free.

Illustrated descriptive Catalogues of all Thompson Bayliss Buildings may be obtained at the Electric Farm or post free from

THOMPSON BAYLISS & CO LTD
DEPT. 68. RAINHAM, ESSEX.
Phone: Rainham 631.

ROYAL SHOW . . CARDIFF
July 5th–9th, 1938

See the
ELECTRIC FARMING DISPLAY
Near the Royal Enclosure and Flower Tents

INTERESTING EXHIBITS AND FREE DEMONSTRATIONS DAILY

THE BRITISH FRAM CONSTRUCTION COMPANY (1911) LTD.
Manufacturers of:—
ARTIFICIAL STONE & CONCRETE PRODUCTS
of all descriptions, including: CABLE COVERS, JOINT BOXES, FIELD GATE POSTS, FENCING POSTS, COPING, STALLS & MANGERS, TROUGHS, and many other items of interest to the Farmer.
Works and Offices:—
Fram Works, WHITCHURCH, Glam.
Telephone: Whitchurch 229.

COMPLETE FARM ON VIEW

THIS year at the Royal Agricultural Show the South Wales and Monmouthshire area committee of the Electrical Development Association, under the chairmanship of Mr. L. Howles, has made arrangements for a farm which is electrically equipped.

It is almost the complete electric farm, with every modern device designed to save labour and money, although the actual layout is dictated by the convenience of visitors as well as the necessities of practical operation.

The farmhouse is modern in design and labour-saving in equipment—fully electrified with an electric cooker, refrigerator, sewing machine, vacuum cleaner, and other gadgets, to give the farmer's wife some leisure for outside interests.

The site chosen for this exhibit is a central one adjacent to the Royal Pavilion and other important buildings, and has as a background a beautiful row of towering trees.

Some idea of the area of the farm will be gained when it is stated that the frontage is approximately 150ft. by a depth of 145ft., the total area being approximately 24,000 sq. ft. The photograph reproduced with this article is of a scale model of the electric farm which was exhibited at the recent successful Welsh Industries Fair, and aroused considerable interest. It gives some idea of the layout of the exhibit.

An endeavour has been made to plan the farm to be labour-saving, and the buildings have been so arranged that the farmer has all the many sections under his easy control.

EARLY PLANTS

Reading the picture from right to left the greenhouse and the cold frames will supply all the farmer's needs in the way of plants for his kitchen garden. The greenhouse and frames are equipped electrically so that the plants can be brought to early maturity.

Next is the small workshop in which the farmer can carry out repair jobs with the aid of electrical machinery.

In the house, which is fully equipped electrically for the farmer's wife, a room has been set aside as the farmer's office. In this office he has control of all the electricity supply on the farm, and is in direct communication by means of loud speakers with each of the sections.

Continued on page twenty-three.

At the 1938 Royal Agricultural Show in Cardiff, Wales, the main feature was "The All Electric Farm", most certainly inspired by the pioneering work of Richard Borlase Matthews on his private experimental farm of Greater Felcourt in Sussex. Thousands of visitors discovered electrical agricultural machinery.

This was almost the complete electric farm with every modern device designed to save labour and money, although the actual layout had been dictated by the convenience of visitors as well as the necessities of practical operation. A farmhouse was fully equipped with electric refrigerator, cooker, sewing machine, vacuum cleaner and other gadgets to give the farmer's wife some leisure for outside interests. A greenhouse and cold frames were heated electrically so that the plants could be brought to early maturity. A small workshop was equipped with electrical machinery for repairs. In the house, the office was fitted outso that the farmer could control all the electricity supply on the farm and was in direct contactwith each of the sections of the farm by means of speakers. Electric trucks provided by Cleco, Metrovick, and Partridge, Wilson & Co., garaged, maintained and charged on the farm, were engaged in transportation between the various farm buildings and for delivery duties to Cardiff retailers and Great Western Railway's railway station. A home dairy was equipped with electric butter makers and cheese grinders. There was a daily demonstration of ice-cream making with electrical equipment and samples freely distributed. The poultry section included 15 m (48ft) long intensive houses with large and small incubators, battery brooders, electrically lit chicken runs and hovers. There was electric machinery, mostly made by the English Electric Company, for grinding corn, chaffing of hay, root slicing and pulping, and cake crushing. The centre two-storey building had an electric hoist for loading and unloading an electric vehicle. Driers were making the corn harvest and hayseed independent of the weather, for the corn and hay could be stacked in one day, the elevator being driven by an electric motor. The Electric Farm was lit throughout by Mazda lamps produced by British Thomson-Houston with their advertising slogan, "They stay brighter, longer."

The local *Western Mail & South West News* published a 40-page supplement about the Show, including a 3-page article entitled "The All-Electric Farm: Wonderful exhibit of the Electrical Development Association' while on Tuesday 5thJuly, 1938 an article by Charles Hubbard was published by the same newspaper, "The Farmer's Talisman. All Operations carried out by Electricity."

His Royal Highness, Prince George, the Duke of Kent paid a visit to the show on 6th July, the second day.

That same year, experiments also took place in Dumfries, Scotland, using electricity in horticulture, grass drying,and even in ploughing.

Borlase Matthews, now in his sixties, continued to campaign. Among his publications "Electricitédans les exploitations agricoles," co-authored with R.H. Drilhon in France.

On the eve of the Second World War, the village of Magnet in the Allier Department in the Auvergne-Rhône-Alpes region (population 650) became the center of an experiment in rural electrification. Here a villager wheels his machine

Illustrations of the electrical farming experiment in the French village of Magnet (pun not intended).

In 1938, to encourage rural electrification, the Intercommunal Union of Electrification of the Allier (Syndicat Intercommunal d'Électrification de l'Allier orsiea), south-east France, directed by Monsieur Joseph Viplé, decided to make the little village of Magnet into a Rural Electrified Commune. For eighteen monthsthe649 inhabitants in 185 houses of Magnet, "les Magnétois", would be provided with electrical appliances, domestic, agricultural or artisanal, necessary for their respective needs, but only having to pay for the cost of the energy consumed, billed at normal rates in the region. The experiment began in August, 1939 a few days before the outbreak of the Second World War which led to the occupation of a large part of Europe by the Germans. France did not escape and was occupied as early as 1940. Mobilization meant thatrural electrification could not be carried out with the planned equipment not being delivered, and also because the supplier, Auvergne Hydroelectric Company had its own problems. Despite the consequences of the Occupation and subsequent difficulties, the Magnet experiment continued ending in December, 1940.

In order to feed their military on the Eastern Front, the Occupier put France under the yoke of rationing making many products rare and almost unavailable. This was the case for petroleum, where vehicle fuel was only available to official users, such as farmers. This did not apply to hydroelectric energy aiding the spread of electricity for a multiplicity of uses, including agriculture.[5]

1942 With the fuel shortage in Occupied France, several companies produced limited numbers of electric vehicles, including this electric cattle truck. (J-N Raymond Collection)

With their men gone off to war, in a besieged Britain, women continued the milk round in their battery electric milk floats (Wythall Transport Museum).

Sadly, Richard Borlase Matthews died of heart failure on 21st August, 1943 at the age of sixty-seven following an unsuccessful attempt to save his youngest son David who had got into difficulties whilst swimming at Port Daifech beach, near Holyhead. Borlase was buried in the grounds of the small chapel at Priory Farm, Ramsdell, near Basingstoke, Hampshire, which was being run by his other son John Borlase Matthews. His grandson, M. J. Matthews, took over the farm in 1950 and it is now run by his grandson, also named Richard Borlase Matthews.

Following the Second World War, French agronomist engineer René Dumont, working alongside Georges Monnet, Minister of Agriculture, championed the modernization of the French countryside, land consolidation and drainage, then the green revolution (that is to say, high-yielding varieties) to feed the Third World and modernize the French countryside. He was an advocate of the use of hydroelectric power and electric ploughing so as to reduce coal imports. However, what was then called 'white coal'was soon abandoned in favour of 'black gold' or petrol. Electric ploughing once again took a back seat.

From 1945 to 1949, the number of electric tractors on Soviet farms increased by 300%, and they worked so well that a further large-scale increase was approved. Photographs in one report showed tracklayer and wheeled versions at work, both linked by cable to an electricity supply. The Soviet electric tractor programme may have been less ambitious than the publicity suggested and later, a senior agricultural official from the USSR said the programme had been abandoned in the early 1950s due to technical problems.Only a small number of tractors were made, and they were just a brief episode in the history of Russian power farming. One of the big problems faced by the electric tractor pioneers was dealing with the power supply cable, and this may have been one of the reasons for the failure of the Soviet programme.

A British attempt to overcome some of these problems was shown in 1949 when the Electrical Research Association, based near Reading, Berkshire, demonstrated a Ransomes MG tracklayer that had been modified by fitting a 9hp electric motor instead of the usual petrol engine.The electricity supply came through a cable connecting the tractor to a power point at the top of a 10.7 m-high pylon (35ft) in the middle of the field. The connector could turn through 360degrees, allowing the tractor to work in ever-decreasing circles, and as it progressed closer to the pylon, counterweights took up the slack to keep the cable under tension. The cable used for the demonstration allowed the tractor to work in a circle with 33.5 m (110ft) maximum radius, which could presumably be increased by using a longer cable, but the rotary work pattern might be a problem on farms that did not have circular fields.

The idea of the electric tractor did not disappear altogether. Nn 1955, *Time* magazine declared Louis Marx of New York the 'Toy King' of the United States. Among his range was a battery-operated reversible diesel electric tractor. Using 2 "D" cell dry batteries, the red and yellow tractor was in plastic and the back wheels were in tin litho. Another Marx model was the battery cable-operated Caterpillar tractor complete with headlights.

Full-scale, attempts were made to launch electrically-operated farm equipment and accessories, some more successful than others. In 1954 International Harvester teamed up with General Electric to launch their *Electrall* system which consisted of a 208V three-phase alternating-current generator connected with electric cables to the device to be powered. The generator could even power a household. A 10 kW *Electrall* generator was an option on the Farmall 400 tractor and a 12.5 kW PTO-driven version was made. The possible applications of *Electrall*power were many, but few made it to market. IH marketing materials showed a haybaler being *Electrall* powered. One of the more novel applications of the *Electrall*was a device to electrocute insects in the field at night, basically like a modern-day bug zapper, but on a larger scale.

In the early 1950s, an English engineer, Professor Thomas Bacon of Cambridge University, was making considerable progress developing the first practical hydrogen–oxygen fuel cell to present large scale demonstrations. One of the first of these demonstrations consisted of a 1959 Allis-Chalmers farm tractor powered by a stack of 1,008 cells. With 15,000 watts of power, the tractor generated enough power to pull a weight of about 1,360 kg (3,000 pounds). On 15th October, 1959, Harry Karl Ihrig, and a team of engineers from the Allis-Chalmers Research Division unveiled a 15kW prototype fuel cell tractor at the company-owned golf course just outside of West Allis, Wisconsin.It was built on the D10/D12tractor chassis but had little resemblance to it. It had a bulky, box-like appearance. Three large panels covered the complex system of fuel cells where an engine would normally be. The operator sat dwarfed behind the giant fuel cell unit. The dash panel was packed with gauges and meters to monitor the chemical process and electric current. To the left of the operator were levers to control the current (for speed) and polarity of the current (for forward or reverse). Oxygen tanks were secured beneath the tractor, and a propane tank was behind the driver seat. It was a one-of-a-kind tractor.That day, the prototype successfully toweda double-bottom plough up and down a field of alfalfa. It remained a one-off.

In 1959, a team led by Harry K. Ihrig at the Allis-Chalmers Company developed and tested this fuel cell tractor prototype at the company-owned golf course just outside of West Allis, Wisconsin. It remained a one-off.

During the 20thcentury, it was almost impossible to compete with mass-produced petrol-engined tractors. John Deere of Waterloo, Iowa produced their Standard-tread 82 litre (18 gallon) *Model D* tractor, of which 160,000 were built and used for ploughing between 1923 and 1953.In 1926, Harry Ferguson applied for a British patent for his three-point hitch, a three-point attachment of the implement to the tractor and the simplest and only statically determinate way of joining two bodies in engineering. The Ferguson-Brown Company produced the *Model A* Ferguson-Brown tractor with a Ferguson-designed hydraulic hitch, of which 1,356 produced. In 1938 Ferguson entered into collaboration with Henry Ford to produce the Ford-Ferguson *9N* tractor. The three-point hitch soon became the favourite hitch attachment system among farmers around the world. International Harvester manufactured the Farmall*Model H* as the row crop version of the McCormick-Deering *W-4*. Between 1939 and 1953, about 391,000 *Model H*s were built. The USSR went to 1st place in the world on the annual production of combine harvesters: 44 thousand units against 29 thousand in the United States. John Deere's *Model 4020* came with a six-cylinder engine, capable of 96 horsepower and their Syncro Range transmission. It was standard in the *4020*s and allowed users to operate at 8 forward speeds and 2 reverse speeds. Between 1963 and 1972, over 184,000 tractors were sold.

During the late 1950s the sintered plate nickel-cadmium batteries became commercially available. In 1961 Robert H "Bob" RileyJr., the in-house inventor for the Black & Decker Manufacturing Company in Towson, Maryland, filed for a patent for a cordless electric drill. It was powered by a slide-out rechargeable NiCad battery pack, wherein the battery pack when discharged, could be readily removed for recharging purposes or else quickly replaced with a spare fully-recharged battery. Bob Riley then adapted the technology to a hedge trimmer. Both these were world firsts. When he retired in 1978, Riley and the B&D team had obtained no less than 22 patents.

In August, 1967 the UK Electric Vehicle Association put out a press release stating that Britain had more battery-electric vehicles on its roads than the rest of the world put together. All manufacturers of battery electric vehicles were, at one time, members of the Electric Vehicle Association of Great Britain, and they received returns from the manufacturers on a regular basis, so they were able to give accurate numbers of BERVs in use in the UK for a certain year. The EVA also had industrial truck manufacturers, battery manufacturers and component suppliers as members of the Association. Closer inspection disclosed that almost all of the 30,000 battery driven vehicles licensed for UK road use were milk floats or door-to-door delivery vehicles, the final link from electric milking machines at the dairy farm.

Of the dozen manufacturers, three indicate the trend. T.H. Lewis Ltd. of Watford had been building milk floats, milk carts and horse-drawn vehicles for London's Express Dairies Company since 1873. In the early 1930s Lewis switched over to electric. They designed two types: a 3-wheeled pedestrian-controlled vehicle with a 178 kg (3.5 cwt) payload, which had a fixed speed of 4.8 km/h (3 mph) on level ground.Midland Electric Vehicles Ltd., established in 1935was based in Leamington Spa, Warwickshire. They supplied chassis to both the Birmingham and Wolverhampton divisions of the Midland Counties Dairy in 1936, and the first production vehicle probably went to the Wolverhampton division, whose ride-on fleet of milk floats consisted almost entirely of Midland vehicles, apart from some bought elsewhere during the war years and some pedestrian-controlled vehicles.A late comer was Wales & Edwards, formerly a garage and car salesroom for Morris and Wolseley cars, based in Shrewsbury. In 1951, Mervyn Morris designed and built an electric milk float for Roddington Dairy. A request from United Dairies saw the production of a 3-wheeled chain driven vehicle, which was an immediate success. An order for 1,500 vehicles followed[6].

By the mid 1960s, the Electricity Council's Research Centre at Capenhurst, near Chester, had a section specialising in agricultural and horticultural problems. When the new Royal Showground at Stoneleigh Warwickshire opened in 1967, it had an Electro Agricultural Centre called the Farm Electric Centre, Stoneleigh, originally run by the nationalised electricity industry to provide information and research services to the farming and horticultural sector. Between 1955 and 1965, the Electricity Development Association published was a series of farm electrification handbooks, substantial booklets of 70 to 100 pages covering such topics as electric water pumps, farm lighting, electricity in the dairy, grain drying, milling and mixing, green-crop drying and automatic feeding.

The General Electric *Elec-Trak* was the first commercially produced all-electric garden tractor, made mostly between 1969 and 1975.The beginning of the *Elec-Trak* really started in 1968 when GE demonstrated an experimental electric car the *Delta* with a top speed of 89 km/h (55 mph) with a range of 64 km (40 miles). It was approximately 3 metres (10 ft) long and was designed to carry two adults and two children. The project was headed by G.E.'s Bruce R. Laumeister. Although the *Delta* was ultimately determined to be impractical, Laumeister realized the very things that made the electric car impractical were benefits to an electric tractor. The weight of the batteries was an enemy of the automobile but was an advantage to a tractor. Limited range was not a problem as garden tractors were rarely used for long periods of time and fewer batteries meant less recharging time.Laumeister concluded that GE should not go into the car business since they did not have the experience, the suspensions, transmissions and whatever else was required to produce electric cars.During 1969, the design of the *Elec-Trak* was worked out by GE engineers led by Laumeister.By September, the project took on the new name of the Outdoor Power Equipment Operation (O.P.E.O.) and was transferred from New Business Development group in the G.E. Research & Development Center to be a stand-alone operating unit of the Transportation Systems Business Division (TSBD) headquartered in Erie, Pennsylvania. In September, 1969 the new operating unit already had 40 employees. In December, 1969 prototypes of the *Elec-Trak* were shown to dealers and customers to test the market.

Laumeister later recalled:

> Indeed, safety was a big concern so the tractor was designed so that when you got off the seat, the motor blades stopped in a tenth of a second. And if you hit the brake, the mower stopped at the same time. We took the tractor to the flower show in New York City to display it. I set it with the front of the mower up on edge so the blade area showed, except we took the blades off and put kaleidoscope discs on instead. And I hired an acrobat who would sit on the tractor, start the mower going, with disks rotating, and he would dive over the front of the tractor to see if he could get to the disks, which would have been the blades, before they stopped rotating and he couldn't. We sold a lot of tractors with that gimmick[7].

The General Electric *Elec-Trak*, developed by a team led by Bruce R. Laumeister in Erie, Pennsylvania, was the first commercially produced all-electric garden tractor, made mostly between 1969 and 1975.

With Laumeister in charge, by early 1970 they had built the entire factory and had models going down the assembly line even though the original appropriation request called for a start up by late 1971! The long-term business plan actually forecasted additional income from an electric sport vehicle. Production of the *Elec-Trak* began in 1971, and 25,000 tractors were produced in 1972. By the end of 1973 the *Elec-Trak* operation had almost 300 employees. The plant had grown from 3716 m^2 (40,000 ft^2) to close to 18,600 m^2 (200,000 ft^2) of production office and warehouse in three different buildings in the Scotia industrial park. With few exceptions such as the mower deck and the tractor deck, all fabrication work was done in house. Those exceptions were due to the fact they did not have large enough presses to make those items. The drive motors were made by the G.E. plant in Fort Wayne, Indiana. Even the circuit boards and wiring harnesses were built in-house.

They nearly all had a yellow livery and several models were produced, including: the *E8M* and *ER8-36* (8 hp); the *E10M (*10 hp); the *E12* and *E12M* (12 hp); the *E12S* and *E15* (14 hp); the *E16* (an upgraded version of the *E15*), and the *E20* (16 hp). The *E8M* and *ER8-36* were styled more as ride-on mowers than tractors. The 'M' suffix used on some models indicates the ability to accommodate a mid-mounted (belly) mower, and an 'H" was used on some models to indicate a heavy duty, double sized battery pack.GE also made an industrial version of the *Elec-Trak*, the *I-5*. It was orange instead of yellow, and had fenders over the front wheels and attachment points for a roll cage and forklifts of varying heights, but was otherwise identical to the *E20*.

Elec-Trak produced branded attachments including electric trimmers, edgers, chainsaws, radios, arc welders, forklifts, front-end loaders, rotary brooms, roller aerators, lawn rollers, dump carts, large vacuums, agricultural sprayers, moldboard ploughs, row crop cultivators, tillers, disk harrows, sickle bar mowers, belly mowers, front-mounted rotary mowers, front or rear-mounted ganged reel mowers, lawn sweepers, electric rakes, snow-blowers, snow-plows, golf bag holders, double seats, 120vac rotary inverters, canopy tops, and more. Most of these attachments were connected by a 36VDC "power take off" using a NEMA 10-50 outlet, usually used in the U.S. for 240VAC clothes dryers and thus compatible with readily available cords in most of the USA.

1973 had seen the Arab oil embargo arrive which should have been a plus for an electric tractor but overall it had caused a major recession, and starting that year all tractor manufacturers had disappointing sales. Compounding the issues was a G.E. union strike in early 1973 just when the factory needed to mass produce units to fill dealer showrooms for the spring season. This was one of the largest strikes G.E. had ever encountered and factory employees honoured the picket line. Salaried staff were required to perform regular duties 4 hours per day and build tractors 6 hours per day. This resulted in unresolved design, manufacturing and quality issues and caused inferior product to be produced by untrained personnel. The result was numerous field service problems that eroded dealer confidence and spread negative advertising to would-be buyers.

In the summer of 1973, G.E. sent in Thomas A.Vanderslice, a hard-line trouble-shooter to review the situation and make recommendations. It is apparent the decision was made to sell off the division as they had already spoken to Simplicity, but they were experiencing the same receivables issues as well as a drop in sales. Plans were made to talk to other potential buyers. Additional moves were made to 'stop the bleeding'. These moves caused the market to predict G.E.'s withdrawal from the marketplace and added more to the problems.

In the spring of 1974 talks began with Wheel Horse and on 9th August, 1974 the deal was done. Wheel Horse ran the operation in Scotia for only a year and then closed it. Wheel Horse and the private brand for New Idea eventually ended all production of the *Elec-Trak* models. Wheel Horse came out with a new model, the *A-60* (1975-1977) and two others, the *E-81* and the *E-141* (1980-1983).

In 1977 James H Downing Jr. of Missouri obtained USPatent 4113044A for :

> A modular farm tractor whose wheels are driven by individual A.C. induction motors and whose accessories or attachments are also driven by A.C. induction motors. A prime mover, such as a gas turbine operating at a constant speed, is mounted on the tractor and drives a three-phase alternator which produces A.C. power at a constant frequency of 400 Hz. The frequency of the power applied to the wheel motors is controlled by a first cycloconverter means connected to the output of the alternator. The alternator output is also coupled through a second cycloconverter means to an A.C. power outlet mounted on the tractor. Various farm accessories associated with the tractor are driven by A.C. induction motors connected to the A.C. outlet. The frequency of the power applied to the accessory motors is controlled by the second cycloconverter means. A material-transferring tunnel runs between the front and rear of the tractor along the center line thereof and between the tractor's wheels.

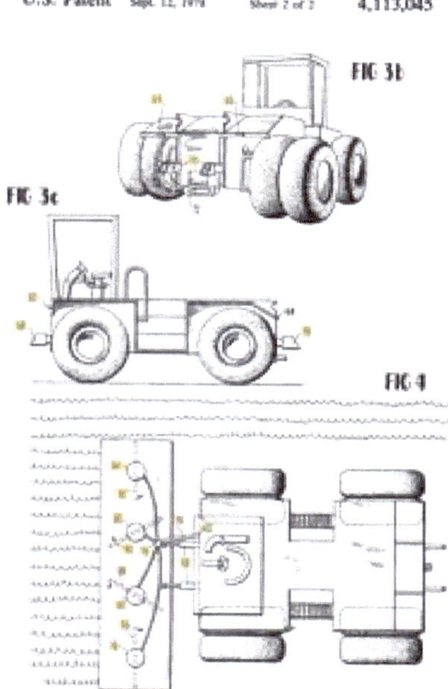

1977 Patent drawing from James A. Downing of Missouri's electric farm tractor prototype with its AC four-wheel drive system. It was never manufactured.

Jim Downing teamed up with Bear Industries, Sedalia, Missouri who were manufacturing gooseneck trailers and other farm equipment, to put what they called the “New World Workhorse” into production. Downing predicted that his all-electric tractor would pull the largest tillage equipment without slipping, thanks to independent induction motors which power the four separate drive wheels. Power flowed to the wheels without differential slippage. Threshing heads, cutters, windrowers, pickers and every other imaginable piece of auxiliary equipment would attach to either the front or back of the unit by way of forklift-like hitches. Truck beds, bins, and tanks of every sort would either be carried or pulled. When pulled, carrying beds could form a land train in which each trailing vehicle was separately powered by its own drive wheels, connected by electric cable. Electric power was fed by simply plugging in to the vehicle ahead. Variable speed control of each pair of wheels, and of each auxiliary implement would be achieved electronically through variation in the frequency of independently supplied alternating current. At all times, the turbine-alternator would run at a constant speed and, while running, never require throttling. An infinite number of speeds, forward or reverse, would be available, while the operating speed of the auxiliary implement or implements would be separately controlled through cyclo-converters.

In short, the mobile unit Downing designed was the basic building block for a complete system of modular agricultural equipment. “The same mobile unit can serve as a power-generating heavy tillage tractor before breakfast, a combine until after lunch, and a heavy duty forage harvester before supper and then a 200 kilowatt standby generator overnight”, Downing explained. “At different seasons of the year, the vehicle could be variously configured as a tractor pulling plough bottoms, as a single-pass planter and chemical applicator, small-grain combine, or as a road-grader or snow-blower.”[8]

Downing had first considered the use of D.C. electric motors and generators, but abandoned them because they were “too bulky, heavy and costly.” He also explored but gave up on hydraulics: “Hydro-static devices provide either high torque with low speeds, or high speeds with low torque, but it’s impossible to obtain high torque over a wide speed range without the use of a transmission.” Needing individual wheel motors with a very broad speed range, Downing also considered steam turbine engines. “They would work but their efficiency with variable speed operation would be terrible.” His solution: A. C. electric power with cyclo-converters, which provide variable speed control of alternating current. In checking with manufacturers of cyclo-converters, Downing was told that while it was theoretically possible to use them to control high-frequency A. C. current on mobile machinery, manufacturers warned that cyclo-converters may be expensive. Undaunted, Downing was satisfied he had figured out a way to make them work economically: "Electric power is the key to making a farm equipment system modular. A modular system is the key to lowering costs of farm machinery and the cycloconverter is the key to mobile electric power." Downing's revolutionary tractor, three decades ahead of its time, was never built.

Another approach was taken by Leslie L. Christianson, Ralph Alcock, Donell P. Froehlich and Mylo A. Hellickson, agricultural engineers at South Dakota State University who started work on their *Choremaster* tractor project in 1983, using two 32-cell battery blocks to produce a 43.5kWh electricity supply powering two motors.One motor operated the hydraulic systems, including the power steering plus the pto, and the other powered a three-range hydrostatic transmission with four-wheel drive.Because of the limitations of battery power in the 1980s, the *Choremaster I* and *II* were designed mainly as yard tractors working within easy reach of a power point for battery charging.The two batteries weighed about 2 tonne (2ton) and it took eight hours to recharge them fully to provide about six hours of work time. The *Choremaster II* was used for several years on the university farm, and the engineers also developed a battery-powered skid-steer loader. US. Patent 4662472A was obtained in 1984.[9]

Another e-tractor pioneer, Steve Heckeroth of Albion, California has been working on alternatives to fossil fuel use in both his private and professional life since 1970 when he was 22 years old. From the first Earth Day in 1970, Stephen D.Heckeroth dedicated his life to finding alternatives to burning non-renewable resources because of the negative effects on future generations. In 1973, he obtained his BA in Architecture at Arizona State University where he planned to pioneer the design of zero energy solar homes with solar water heaters and solar radiant floors. During the 1970s Heckeroth taught classes in sustainability and renewable energy at College of the Redwoods, Fort Bragg, California. In 1991, he also designed the 279 m^2 (3,000 ft^2) Caspar Point home, located in northern California on a rugged point of land jutting out into the Pacific. It incorporates southern exposure, sun spaces, thermal mass, insulating envelope, earth coupling, and thermosiphon solar-heated water for domestic hot water and radiant floors. He has designed and built over 20 such passive solar homes.

But Heckeroth realised that transportation uses four times as much energy as housing and causes ten times more air pollution. Encouraged by the California Air Resources Board (CARB) Zero emission mandate of 1990,with off-the-shelf golf cart and forklift technology, Steve started converting every vehicle he could get his hands on. In a search for the lightest possible vehicle, he found fibreglass Porsche *550 Spyder* replicas that weighed less than 445 kg (1000 lb) but had a range of 161 km (100 miles). Mendocino is a coastal community in northern California. The traditional wooden water towers of the local Mendocino area gave the inspiration to Steve when he designed and built his own Tower House. In 1992, he founded MendoMotive in Fort Bragg, California to develop and build EVs. Understanding that the weight of lead acid batteries was the biggest problem for electric automobiles, Heckeroth switched his attention to farm tractors because they need weight for traction. He also found that electric motors have the instant torque at low speed that makes tractors ideal electric vehicles. Following a first agricultural electric tractor prototype, in 1993 he built an electric rototiller and converted small agricultural tractor to battery AC electric propulsion, going on to build a number of one-offs for oil independence and zero emissions on small farms and vineyards. In 1995, commissioned by Ford-New Holland, MendoMotive built an electric agricultural tractor with front loader. A 1996 version followed with motors in its wheels. In 1997, Heckeroth designed and built three full function prototype electric tractors to clear explosives from farm land after wars for a Japanese manufacturer. These incorporated innovations such assteer-by-wire zero radius steering, remote control operation, on-board inverter/chargers for mobile AC power, motors mounted on implements to replace dangerous pto's and the first ever use of linear actuators on an electric tractor. Continuing to innovate, Heckeroth built a tractor with a solar shade canopy, another for remote-control and in 1998, for the Agricultural College mechanical engineering programme, an electric tractor wire-controlled by joystick, the first electric metal track loader (2000) and the first all-electric rubber track crawler (2006).

Heckeroth was not alone. In 1994, German machinery dealer Roland Schmetz developed the diesel-electric tractor, based on a New Holland *M135* which he called the *Eltrac E135*.[10]

In 1994, the Chinese took out a patent for a half-feeding type threshing machine which could use solar power to thresh rice and wheat. Electric energy generated by a solar energy system in the utility model would use a motor to drive a double curve type threshing rolling cylinder to carry out threshing work. Simultaneously, a wind wheel would be driven to blow a winnowing chamber, the threshed rice and wheat winnowed in the winnowing chamber, impurities after the winnowing is exhausted by an impurity discharge port. The cleaned rice and wheat is directly put into a basket or a bag through a rice and wheat outlet.

These were the vanguard of a transition to take electric agriculture machinery into a new and finally more successful period.

Heckeroth Electric Tractor (ET) Innovations '92

First Electric Full Function Agricultural ET '92

First ET with Mobile AC Power '93

First Electric Agricultural Implement '93

First ET Built for Major Tractor Manufacturer (Ford/New Holland) '95

First Electric Tractor with Wheel Motor '96

First Tractor with Separate Electric PTO Motor '96

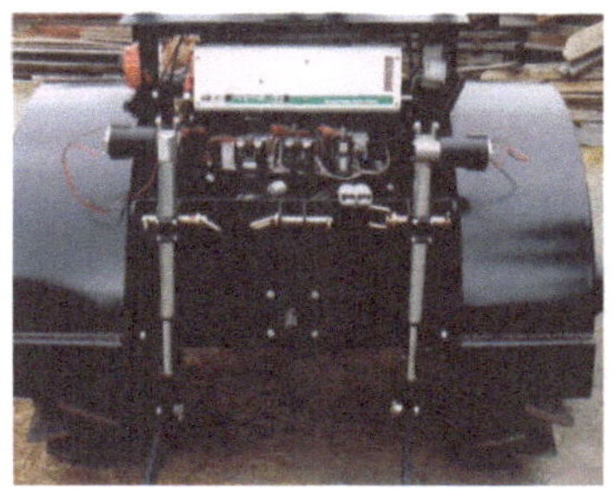

First Tractorto to use Linear Actuators on Rear 3 Point Hitch '96

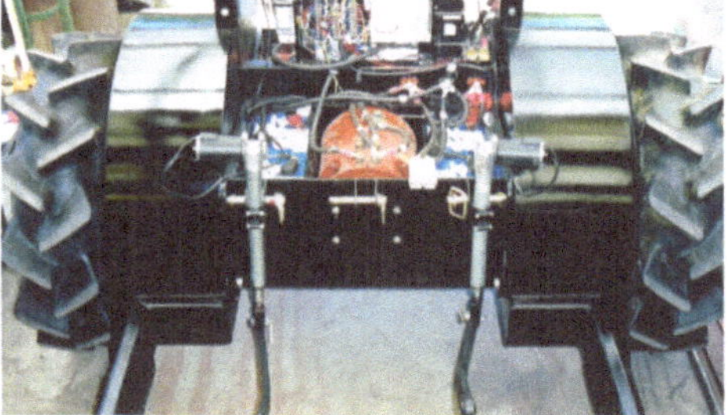

First Tractorto use Linear Actuators on Mid 3 Point Hitch '96

During the 1990s Stephen D. Heckeroth of MendoMotive in Fort Bragg, California innovated a whole range of electric tractors (*Courtesy:* Stephen Heckeroth).

First ET with Exchangeable Battery Pack '96

First ET with Zero Radius Steering '96

First Tractor with Solar Shade Canopy '96

First Electric Tractor with Reversible Seat '96

First ET with Seat in Planting / Harvesting Position '96

First Remote Control ET '97

First Electric MetalTrack Crawler 2007

First Electric Rubber Track Crawler 2008

First ET with Mid Hitch for Cultivation 2013

First ET with Front, Mid and Rear Hitch 2015

Endnotes

[1]Kevin Desmond, *Electric Trucks: A History of Delivery Vehicles, Semis, Forklifts and Others* (Jefferson, NC: McFarland, 2019).

[2]Stephonie J.Clarke, "The All Electric Farm at Greater Felcourt. " Felbridge & District History Group, July 2013.

[3]"The use of electricity in agriculture being a report of a conference held at Rothamsted on January 29th 1936: F.E. Rowland "Electric Motors for Farm Machinery.

[4]G. H. Cashen, M.Sc., and. Dr. B. A. Keen, F.R.S, "Measurements Comparing Fuel and Electricity as a Source of Power" Rothamsted Experimental Station, 1937.

[5]Jacques Sorbets, L'électrification des campagnes L'Illustration, n°5114, 15 mars 1941.

[6] Kevin Desmond, Electric Trucks, A History of Delivery Vehicles, Semis, Forklifts and Others, Jefferson, NC: McFarland, 2019.

[7]Alan Ginsburg, "Enthusiasts repair, restore models of Elec-Trak invented by GE engineer," The Daily Gazette, April 11th, 2019.

[8]All-Electric Tractor Being Developed," Farm Show, 1977, Volume 1, Issue 2, page 12.

[9]Michael Williams, "Machinery Milestones: Electric tractor power," Farmers Weekly, June 21, 2019.

[10]Club of Bologna 21st Annual Meeting Key Note Report Bologna, EIMA International, Nov 13-14, 2010.

3

The Last Twenty Years (2000 to 2020)

The 21st century dawned with interest in the potential of electric agricultural machines. Beginning in 2000, a concept study, MELA Mobile Elektrische Leistungs-und Antriebstechnik Mobile (electric power train technology)supported by universities and organizations in Germany as well as Fendt Agricutural Company (AGCO) showed a possible solution on tractor electrification with electric driven auxiliaries as well as electric CVTs and implement electrification. In 2003 the University of Hohenheim developed an electric combine harvester called *Magda*. With the introduction of a 42 VDC power supply system, the range of small rotational drives and servo motors could be covered. They showed that if more power had to be transmitted, three-phase alternating current should be used since three-phase machines had a large speed range and a large range of constant power.

During the first years of the century, long-established transport manufacturers have been quick to see the environmental advantage of electric propulsion, be it with batteries, hybrid-electric or fuel cell. With airliners, Boeing, Airbus and Tupolev quickly saw the advantage. With motorcycles, names such as Harley-Davidson, Curtiss, Ducati, Saroléa and Royal Enfield have positioned battery packs between two wheels. With trucks, Daimler, Toyota, Cummins, Peterbilt, DAF, Volvo, Leyland, Kenworth, and Isuzi, among others have taken the electric route. With buses, Alexander Dennis, Scania, Volvo, New Flyer, Van Hool, DAF-VDL, Blue Bird, Heuliez, Rampinican carry their passengers more silently. It therefore seemed only logical that traditional agricultural machinery builders should make the same transition. The first to do this was John Deere, established in Moline, Illinois in 1837.

Deere's ePower Technologies Group led by Bruce E. Wood, began its transition with a prototype gasoline-electric zero-turning radius lawn tractor and its currently available electric-powered *E-Gator* turf utility vehicle. Then, in 2002 Deere teamed up with fuel cell manufacturer Hydrogenics Corp., Mississauga, Ontario, Canada, using Hydrogenics' third-generation HyPM-LP2 fuel-cell power module, a 5,000-psi lightweight tank from Canada's Dynetek, to develop to develop a technology demonstrator fuel cell-powered Commercial Work Vehicle (CWV).On 20th May, 2003 Bruce Wood gave a talk at the 9th National Clean Cities Conference & Exposition in Palm Springs, entitled "Perspectives on Lowering Diesel Emissions", in which he presented the CWV prototype. With a Solectria drivetrain, the CWV had four-wheel drive and steering, and full drive-by-wire capability that allowed the vehicle to be driven by remote control or robotically (steering encoder by the international SKF Group). It had a power strip on the back which drew power from the fuel cell, allowing the CWV to act as a 20-kilowatt generator. A diesel-electric version of the same vehicle joined the U.S. National Army Corps.

In 2000, John Deere Electronic Solutions was joined by Dr Noel W. Anderson, long-time resident of Fargo, North Dakota who had been working in in the areas of precision agriculture, autonomous off-road equipment, and UAVs for over 25 years, having one hundred-plus patents in these fields. These include a grain yield monitor, a soil sampler, variable rateplanting and seeding, variable rate fertilizer application, and mapping software. In 2006, Anderson obtained a key patent for an agricultural automation system with field robot, to which were added such innovations as a method for governing the speed of an autonomous vehicle.

Case, founded in 1842 in Racine, Wisconsin, becoming Case IH in 1985, was the first to make a diesel-powered tractor. In 2005, Case IH modified one of their *MXM* tractors to a diesel-electric. The ProHybrid *EECVT*, combining a 120 kW (160 hp) diesel engine offered 800 Nm torque with two 50kW (67 hp) electric motor/generators. The prototype drew on technology from the Steyr continuously variable transmission (CVT) now in Case's *CVX* tractors. One of the motors acted as a generator providing electrical power to the other which served as a traction motor. Excess energy was stored in a large 456V, 11.5 kWh battery that sat on the tractor's nose in place of a front weight. Regenerative braking recharged the battery. In addition to an all-electric mode operation, the system could channel drive from both electric motors and the diesel engine through the transmission as required.

As Thomas Edison had understood back in the 1900s, if the battery was right, everything would fall into place for all forms of electric transport, including agriculture. He had invested millions of dollars into developing and promoting his nickel-iron battery for cars, trucks and buses, but failed. Although the nickel-cadmium battery and its sister the nickel-metal-hydride had been more promising than the lead-acid, the grail discovery came in the increased development and commercialisation of the lithium ion battery.

In the fall of 1972, M. Stanley Whittingham, an English-born chemist working with a team at the Exxon Research & Engineering Company announced that they had come up with a new battery, and patents were filed within a year. Within a couple of years the parent company Exxon Enterprises wheeled out a 3W 45Ah prototype lithium cell, and linking it to a diesel engine started work on hybrid vehicles. When solid-state physicist, Professor John Goodenough, became head of Inorganic Chemistry at Oxford University in 1976, his research group includedDr. Phil Wiseman, Dr. Koichi Mizushima and Dr. Phil Jones. They set themselves the task of looking at the potential of rechargeable batteries which they began by simply "kicking around ideas on a blackboard". "We looked at it in a different way, using lithium cobalt oxide at the positive terminal and pulling the lithium out; this produced a huge cell voltage, twice that of the Exxon battery,"Dr. Wiseman explained. "It was this spare voltage that allowed alternatives at the other terminal where Exxon had been forced to use lithium metal which was fraught with problems. Instead lithium-ion material could compose both electrodes." The group's research was published in the *Materials Research Bulletin* in 1980. In 1977, Whittingham teamed up with John B. Goodenough to publish a book "Solid State Chemistry of Energy Conversion and Storage."[1]

In 1976 Congress had made a large appropriation for electric vehicle research but as soon as oil prices fell in the 1980s, companies and governments shut down all those battery-research projects they had started in the 1970s. Exxon closed Whittingham's shop, appointing him director of their chemical engineering division, responsible for technology, for synthetic fuels, chemical plants, and refineries. When Exxon stopped their R&D, the U.S. government largely got out of clean-energy research as well. If President Reagan had continued the programmes the Carter administration started we'd be a lot further ahead. But that didn't happen, If President Reagan had continued the programmes the Carter administration started research would be a lot further ahead. But that did not happen and there was a hiatus.

In 1985, Akiro Yoshino of the Asahi Kasei Corporation in Japan created the commercial prototype lithium-ion battery. Two years later, Sony of Japan entered an agreement with Eveready USA to develop mass market rechargeable lithium batteries. Many alternative cathode and anode chemistries have been discovered since those first commercial Li-ion batteries hit the market in 1991. The original cobalt oxide cathode's vulnerability to overheating, producing oxygen and possibly catching on fire led to the 2006 "era of flaming laptops". Meanwhile, other cathode types such as lithium manganese oxide (LMO) and lithium iron phosphate (LPO) were developed, offering greater resistance to overheating but having less energy density (measured in kilowatt-hours per kilogram, or kWh/kg). These types have become the principal players in the electric vehicle field. Indeed for their achievement in 2019, Professors Goodenough (aged 97) Whittingham and Yoshino jointly received the Nobel Prize for Chemistry.

Development continued and by the late 1990s Japanese companies, in particular Sony, had made great strides in the commercialization of lithium rechargeable batteries. In 1997, TsuyonobuHatazawa, R & D Manager at the Sony Corporation, Kanagawa, invented the polymer gel electrolyte and lithium-ion polymer battery. Also in 1997, Nissan Motors produced its *Altra* , an electric car equipped with a neodymium magnet 62 kW electric motor and run on lithium-ion batteries manufactured by Sony. The following year, Ji Joon Kong in South Korea founded Kokam to manufacture polymer processing equipment, including polyester film and polarized film manufacturing systems, and breathable (porous) film casting systems. In the late '90s, Kokam expanded its business to designing and manufacturing li-ion secondary batteries and succeeded in developing the world's first commercial products. The same year, tests were carried out in France of vehicles equipped with li-ion batteries where an encouraging autonomy of 200 km (124 mi) was observed by such authorities as the Inter-ministerial Committee for Clean Vehicles[2].

During the first decade of the 21stcentury, well managed li-ion batteries became commercially available for other transport systems. In 2007,Torqeedo of Starnberg, Germany were building up their range of electric outboard engines with an integrated li-ion battery pack. The same year, pioneers in the USA, France, Germany and Slovenia built and flew their first electric prototypes using li-ion batteries and Zero Motorcycles of Santa Cruz, California were pioneering the safe use of li-ion batteries in their two-wheelers. Also in the same year Smith Electric Vehicles of Gateshead, England launched their li-ion trucks, the *Ampère*, the *Faraday* and the *Newton* and BAE Systems Inc. unveiled their li-ion system for hybrid buses: so it was for electric agricultural machines, on land, and in the air (see Chapter 4).

Since 1981, Jacques Kremer had run his vineyard in Venteuil in the heart of the Champagne country in France, where he continued to innovate with EAMs such as an electric backpack and an electric wheelbarrow. In February 2009, with •700,000 investment, 5,000 hours of study, a workforce of two to six employees, and a partnership with the champagne house Moët &Chandon, which accompaniedthe development of the machine to meet the needs of wine growers, Kremer developed his *T3E.* This is a straddle grape harvester was powered by 80 kwh lithium-ion-iron-phosphate batteries(LiFePO4) supplied by E4V in Le Mans. For the *T3E,*Kremer was awarded First Prize for Innovation at the VITEFF (Le salon international des techniques des vinseffervescents) Exhibition in Reims. In 2010, this was modified into the 110 bhp (80kW) *T4E*, a four-wheeler more adapted for the whole market of the narrow spaced vineyards.In November, 2013 for the *T4E*, the new company Kremer Energie, based in Beaune, received a silver medal of the Innovation Price at the SITEVI (one of the biggest International exhibitions for the vine-wine, fruit-vegetable sectors and olive growing). The use of electrical equipment enables optimisng the performance of tools and their efficiency. Kremer Energie is working with different manufacturing partners for the development of specific electric tools for the *T4E.*To-date, Kremer has sold 34 units.

Over in the USA, a growing number of farmers decided to electrify the reliable Allis-Chalmers *Model G* tractor, of which some 30,000 units were built between 1948 and 1955. The initiative was led by organic farmer Ron Khosla in New Paltz, New York state and has been followed by over 100 farmers who have found their electric tractors more reliable, quieter, and less smelly than the original gasoline-powered version. Although many of them used lead-acid batteries from 2009, a number of them spent their money on LiFePO4 batteries.

Also in 2009, Stihl, world famous for their saws and trimmers, embraced the li-ion battery and began to produce a range of 36 V cordless machines.

Since 1995, Agritechnica, the largest European exhibition of agricultural machinery, has been held in Hanover, Germany. This fair is organized by the DLG (Deutsche Landwirtschafts-Gesellschaft) - German Agricultural Society. Since 1950, Belarus have manufactured a series of four-wheeled tractors produced at Minsk Tractor Works, MTZ in Minsk, Belarus.As of 2005, it had nearly 20,000 workers, its factory producing over 62 vehiclemodels. At Agritechnica 2009, Belarus showed their *3023* hybrid tractor, with a 220 kW diesel engine and a 172 kW generator, a diesel-electric drive train and an electrically driven front pto (power take off) This tractor gained a DLG medal representing the first agricultural tractor with diesel-electric drive train in series production. Only having an electric power transfer between the engine and the transmission, this arrangement is also suited to be operated as a stationary source for electric power, a 172 kW uninterruptable power supply for remote areas. Individuals were also taking the same approach. Malcolm Lucas from Reeves Plains, South Australia converted an East German Forstschritt straw worker as a base and converted into a diesel-electric harvester: all major moving components were driven by 20 electric motors with a 300 kva generator on top of the harvester.

In 2010, New Holland Agriculture (founded in 1895) announced the establishment in VenariaRealeon the outskirts of Turin, Italy, of the first pilot Energy Independent Farm equipped with New Holland *NH2*™ tractor. Called "La Bellotta", the Energy Independent Farm centred on the capacity of farms to produce electrical energy from renewable sources with low environmental impact, storing it in the form of hydrogen and reusing it easily. The highly innovative content of the project from a technological, energy and environmental standpoint, as well as its full compliance with EU targets for renewable energy, earned it a place among the top projects in the Industria 2015 programme "New technologies for Made in Italy", sponsored by the Italian Ministry for Economic Development. The *NH2*™ hydrogen-powered tractor uses fuel cell technology to generate electricity that powers its electric drive motor, on-board auxiliary apparatus and even electrical tilling implements. Based on the New Holland *T6* tractor, the *6.140*model in operation at La Bellotta will have three fuel cells with a total power output of 100 kWwith continuous torque of 950Nm and maximum torque of 1,200Nm. Top crankshaft speed is 3000 rpm, and efficiency at maximum power output was a staggering 96%. Under the project, various methods of producing the hydrogen that fuels the *NH2*™ tractor will be evaluated, including electricity produced from renewable sources to the most innovative 'reduced' scale steam methane reforming process. It must operate all the implements required for different seasonal operations: soil preparation, seeding, baling, transport, and front loader applications.To facilitate operation of the *NH2*™ tractor, the farm will be equipped with a hydrogen storage tank and a special filling station complete with compressor.

Founded in 1911, Merlo is based in Cuneo, Piemonte. One hundred years later 2012, they produced their *TurboFarmer 40.7* Hybrid, a telescopic handler powered by a downsized 56 kW diesel engine flanked by an electric motor working off a 30 kWh pack of lithium batteries giving at least two hours of full working capacity when using only electric power. The Agritechnica jury of experts reckoned that savings on fuel in terms of the hours and tasks to be handled would allow the extra expenditure to be amortised in "a couple of years". Electrical power would also be very advantageous in confined spaces such as greenhouses, stables, and food production plants etc. since no exhaust fumes are produced and the noise is minimal.

In 2011, a team of arable farmers led by Paul van Ham, an environmentalist electronics engineer based inWageningen, Gelderland Province, the Netherlands, got together to produce a sustainable tractor. Five years later, with fourpatents in the Netherlands and in the EU, with international patents pending, the Multi Tool Trac which appeared in 2017 has a complete electric powertrain with 4 x 22 kW nominal and 4 x 44 kW maximal battery capacity built with reliable components, whilea modern 6-cylinder diesel engine with 160 kW (210 hp) is used as range extender. It has a unique on-the-fly track width adjustment and five positions for all common tools and agricultural equipment. This enables a Controlled Traffic Farming (CTF) system, i.e. keeping on the same tracks resulting in better soils, lower costs and better yields. *Multi Tool Trac*also has three full positions for hitch and pto. It has a continuously adjustable cabin over the whole length of the frame track width on-the-fly adjustable from 2.25 to 3.25 metre (7.4 to 10.7 ft) wheel base 5.50 m (18 ft) loading capacity 5 tonne (5 ton).

In 2014, Poly-lab.net, a consortium of 14 German design and production companies and two research organisations developed the *Kulan* to show what agriculture might look like in 20 years. It is powered by a pair of 2kW electric motors, one in each rear-wheel hub which receive power from 16 li-ionbatteries, producing 48 volts total, located beneath the electric utility vehicle's cargo platform; the batteries are rechargeable via any 220-volt electrical outlet. Neither a car nor a tractor, the *Kulan* can transport things a farm needs on a small or medium-size farm, such as animal feed or harvested fruit. And since there is no noise or exhaust fumes, the *Kulan* does not disturb farm animals. It has a range of nearly 300 km (85 mi) or operate for six hours, depending on variables such as cargo weight and temperature. It has a top speed of50 km/h (31mph).

In 2015, Fendt, founded in 1930 in Marktoberdorf, Bavaria unveiled their *Former 12555 X*electric four-rotor hay rake with a working width of 12.5 m (41 ft) and 12 double tines. It is supplied with electricity by a 700V direct current interface specified by the AEF (Agricultural Industry Electronics Foundation) and has an integral electric drive for each rotor. Fendt has also developed an efficient torque motor for the *Former 12555 X* integrated in each rake socket so separate housing, bearings and gearbox are not required.Every motor is regulated independently via its own power electronics.

In 2015, with his wife Laure, Alexandre Prévault, a 29-year-old farmer and engineer of the Gerzatcommune in the Puy-de-Dôme department in Auvergne in central Franceformed a company,SabiAgri, and built a joystick-controlled electric tractor called the *Alpo* with eight hours autonomy and two-hour recharge time. The batteries are built into the chassis, on the sides, and the engine is in the rear wheels.

Since 1960, the Weidmann brothers of Diemelsee-Flechtdorf, in Hesse, Germany have produced more than 65,000 machines and from 1978 the successful *Hoftrac* articulated wheel-loader for moving around hay bales or compost. In 2017, Weideman electrified with the *eHoftrac*, offering two powers (240 or 310 Ah) to assist the farmer in small daily work over 2 to 5 hours. In addition, Weidmann partnered with Bema to present an electrically powered sweeper/brush accessory.An electric cable thus replaces the hydraulic hoses, running from the machine itself to the front attachment tool.

Since 1995, Georg Mayer of Siloking has been producing trailered vertical feed mixers, self-propelled vertical feed mixers, electrically driven vedrtical feed mixers, stationary vertical mixing and dosing units as well as silage unloading and distribution equipment. His 370-strong workforce has made units exported to over 50 countries around the world. In 2018 Siloking electrified with their *eTruck*, a 100% electric self-propelled feed mixer which comes in three models of 8, 10 and 14 m^3 (283, 353 and 494 ft^3). 15 kW are used for the wheels and 18 kW for the screws with rotation speeds of 17, 33 or 50 rpm. The road speed reaches 20 km/h (12 mph). The electric mixers are accompanied by a self-propelled unloader, also electric.

For mowing and brush cutting the ditches and banks which flank country roads and farmland, Rousseau, established in Lyon, France in 1962, have changed the power transmission of their rotors from hydraulic to electric. Beginning with the *E-Kastor* in 2017, they followed with the *E-Thenor* and the *E-Fulgor* in 2018 and the following year the *E-Xtra 160*, which has a horizontal reach of 2.65 m (8.7 ft) for a rotor width of 1.60 m (5.3 ft). Its working angle is between + 90 and -70 degrees.

From 2011, funded by the European Union and the State of Saxony, Sepp Knüsel of the Rigitrac concern in Küssnacht, Switzerland began to work with the Technical University of Dresden, the Department of Agricultural Systems Technology, EAAT GmbH Chemnitz anda specialist electronic engineering firm. Their goal wasto develop a more manoeuvrable diesel-electric tractor. They understood that electric wheel motors offer many more possibilities in vehicle design because the wheels no longer require drive shafts for power transmission. The connection of the wheels to a tractor body can be implemented with much more design freedom, since the wheels hang in principle only on cables that transmit power for the driving force. The Rigitrac *EWD 120* they produced has four electric single wheel drives with a power of 33 kW per motor. Based on this, they progressed to the Rigitrac *SKE 50 Electric*. With a rated power output of 50kW (68hp), energy comes from an 80 kWh lithium-ion battery giving it 5 hours on a single charge, depending on the nature of the work it is doing. *SKE 50* has five electric motors,the fifth motor powering the hydraulic system. In addition, the tractor has a regenerative braking system. In 2018,Sepp Knüsel and Rigitrac received the "Swiss Innovation Award." Since 2019, the *SKE 50*has been providing municipal and agricultural services on factory yards and farms. All performance data is transmitted online to the Rigitrac research and development department.

In early 2017, John Deere Labs unveiled its fully electric tractor, called the SESAM (Sustainable Energy Supply for Agricultural Machinery) at the Paris International Agribusiness Show. It has a range of about 55 kilometres (34 miles), and it takes about three hours to charge. This was following their acquiring Blue River Technology, a startup with computer vision and machine learning technology that can identify weeds, making it possible to spray herbicides only where they are needed. The technology reduces chemical use by about 95% while also improving yield. This is one step in John Deere's embrace of "precision agriculture," the use of technology to target crops and soil for optimum productivity and health.

During this time, to pay the bills, from 1999 to 2007 backyard electric tractor builder Steve Heckeroth of Albion, California had been working as the Director of Building Integrated Photovoltaics for the largest thin-film solar manufacture in the world designing PV roofing products. After a hostile take-over and closure of the solar company by Chevron, Heckeroth decided to go back to designing and building electric tractors full time. Among his innovations he worked out and patented an efficient battery exchange system. At first, he preferred to remain with lead-acid batteries, arguing that the additional weight of the lead batteries, which weigh 30 kg (600 pounds) each, is helpful on a farm tractor because it provides more traction. The body is 102 cm (40 in) wide and the total outside width, including the wheels, is 122 cm (48 in). He uses lead acid batteries, primarily because lithium batteries are too expensive. A lithium battery costs about $14,000, plus another $2,000 for a battery management system. Lead batteries cost about $1,000 and the tractor requires three to have the same performance as a lithium battery.

Heckeroth believed that the way to stimulate more use of electric tractors would be to scale up production and reduce the selling price so the economics became more favourable. Typically, growers are not charged up front but pay after -harvest when they have the money. In 2012, he received $450,000 grant from the Indo-US Science and Technology Fund (IUSSTF) in New Delhi, India to produce 4 prototype full function electric agricultural tractors.

In 2013, Heckeroth founded MendoMotive's sister company Solectrac LLC in Albion California to take electric tractors into commercial production. Among the continuing innovations from the company are the first electric tractor with mid-hitch for cultivation and the first electric tractor with front, mid and rear hitch (2015).

In 2016, aged 68, the quiet Californian received a $250,000 first phase grant from the National Science Foundation (NSF) to refine key electrical components to help take electric tractors into commercial production. The NSF grant helped identify the best components from the rapidly evolving battery and electric drive industries. The resulting production models are the *eFarmer* and the *eUtility*. As part of the collaboration with the Indian partners, Solectrac is in the process of becoming a distributor for a Compact Electric Tractor (CET).

Solectrac's *eFarmer* is designed and built from scratch for cultivating row crops and performing most other tasks on small farms and provides the equivalent of 30 hp. It has 46 cm (18in) of clearance, adjustable track width from 112 cm to 173 cm (44-68 in) and three parallel acting hitches. The front hitch accommodates a loader, reaper, or exchangeable battery pack to balance the weight of an implement on the back. The mid hitch allows forward visibility for precision row crop cultivation and can also be used to exchange auxiliary battery packs. The rear hitch and Power Take Off (PTO) are for all rear-mounted implements or exchangeable battery pack to balance the weight of a front mounted loader or implement. Solectrac tractors take full advantage of the 272 kg (600pound) onboard and exchangeable battery packs to increase traction and balance the weight of implements. Each tractor has a safe 48-volt, 20-kWh lithium iron phosphate (LFP) onboard battery pack. Patented exchangeable battery packs can be quickly swapped using the tractors' three-point hitches to allow extended or continuous operation. The battery packs may be charged from the electric grid or from renewable sources of energy such as solar, wind or micro-hydro. When charged from onsite renewable energy, farmers can be grid independent and eliminate all upstream emissions.

Heckeroth's philosophy:

> Solectrac is currently in the final stages of development with demonstration and testing underway. Commercial production is slated for Q3 2019. Reaching this milestone is a culmination of the last 25 years of experience. It is my opinion that we are at the point where we need to start focusing on what is necessary for survival on a changing planet. Life in the oceans in particular is in steep decline. Oxygen water and food are the perquisites for life. So I have chosen to work on the production of food in the most efficient way while slowing down the production of GHG emissions. Our electric tractors are 2,000 times less noisy, produce zero greenhouse gas emissions and particulates in the field, use 8 times less energy and have 10 times less maintenance when compared with similar sized diesel tractors. The electric motors have only one moving part and are 95% efficient. They offer instant and maximum torque for all the low speed tasks performed by tractors. In addition, electric motors do not need to idle. Diesel engines have more than 300 moving parts, are less than 30% efficient, require regular expensive maintenance and idle even when work is not being done.[3]

Over in Switzerland, Sepp Knüsel and his four daughters (Theres, Ruth Edith and Doris) at Küssnacht am Rigihave built the RigiTrac *SKE 50* electric tractor. With a rated power output of 50kW (68hp). The tractor runs off an 80kWh lithium-ion battery. It can run for up to five hours on a single charge, depending on the nature of the work it is doing. The *SKE 50* has five electric motors. Each PTO (front and rear) and axle has its own dedicated motor. The fifth motor powers the hydraulic system. This year, the tractor will reportedly undergo evaluation. Likely applications will include tasks on small farms and in urban environments. The front axle is home to 425/55 R17 tyres; the rear is fitted with 440/65 R24 tyres. The tractor (unladen) weighs only 2.8tonne (2.8 ton).

Fendt in Marktobernorf, Germany, unveiled their e100 Vario battery-electric compact tractor with an output of 68 hp (50 kW) and five hours of operation per charge. The e100 Vario runs a 650 V lithium-ion battery with a capacity of 100 kWh, which can be recharged up to 80 percent in 40 minutes (Fendt).

Using an IEC 62196 Type 2 connector, which is the standard for charging electric cars in Europe (and perhaps in America too in the near future). The *e100 Vario* is fully compatible with existing hydraulic implements and it also has two AEF-compliant power interfaces for electrical equipment, with a short-term boost available of up to 150 kW. Air conditioning of the tractor's cab (and the battery and electronics) is achieved with an energy-efficient, regulated, electrical heat pump and because the *e100 Vario* connects with a smartphone, the cab temperature can be warmed or cooled in advance and charging status can be monitored remotely.

On 1st April, 2018, Porsche Newsroom announced their electric tractor:

> The tractor joins the rest of Porsche's *Mission E* electric vehicle line-up, which includes the electric *Mission E* sedan to defeat the Tesla *Model S* and the *Mission E Cross Turismo*, an amalgamation of the *Mission E* sedan and a lifted wagon like an Audi *Allroad*. This bold vision for 21st century agriculture blends design cues from Porsche's original mid 1950s tractors with the same advanced digital connectivity and 800v fast-charging architecture that powers the *Mission E*. *Dear reader, the editorial team has taken the liberty of having a little fun here on April 1. Of course, the Mission E tractor will not be built.*

Turkey has also begun to build electric tractors. An Azerbaijan-Belarus enterprise for the assembly of tractors of the Minsk Tractor Factory (MTZ) has been established in the Turkish city Kirikkale with the support of the Ganja Automobile Plant, which has been cooperating with MTZ for many years.A special version of the Belarus *1221* model of the Euro-4 environmental standard has been developed for the Turkish market. The plan is to produce 100 e-tractors in 2019, gradually increasing production to 2,000 units per year.

In 2019, the Carraro family business of Campodesega in northern Italy, making vineyard tractors since the 1980s, presented their *Ibrido* tractor, enabling farmers to choose whether they wish to work inpure electric mode for use inside closed structures, greenhouses and stables; or for municipal applications, in diesel only for road transportation or when not carrying/towing heavy loads; or as a hybrid for transport operations when towing or carrying loads, or for heavy PTO operations. With the electric motor offering up to 105hp, the same output as the diesel engine, the tractor can be used on a standalone electric basis, or in conjunction with the original engine. With a driveline that still retains the original 24F/24R speed layout, irrespective of the chosen power source, the set-up should provide impressive torque response given the nature of electric motors.

In May, 2019 Volvo Construction Equipment unveiled two zero-emission electric compact machines at the Bauma show in Germany,the *ECR25* excavator and the *L25* wheel loader. These machines were the first to be shown from a new electric range of Volvo-branded compact excavators and compact wheel loaders. The *L25* wheel loader is powered by lithium-ion batteries which allow for eight hours of operation for general work. It also has two electric motors, one for the drivetrain and one for the hydraulics.

In November, 2017 the Escorts Group, founded in 1960 in Haryana, India aiming to cater for the fruit, horticulture and homestead markets presented the Farmtrac 4WD *Electric 26E*, India's first electric-powered tractor, unveiled at Agritechnica. Soon after, Escorts formed a joint venture with Japan's Kubota Corporation to produce high-end technology and new-age tractors for the growing demands of highly-mechanised farming. In 2019, under the guidance of Professor Prashanth Nayak, Sayed Rouhan Rawaha, Mohammed Jasim and Jamal K, Mahipal Singh, students of Srinivas Institute of Technology (SIT), Valachilin the Indian state of Karnataka have created *Farm Runner*, a multipurpose 1hp electric tractor able to plough, sow seed and to harvest. The cutter is placed at front part of the vehicle for harvesting and plough and drum for sowing the seeds are placed at the rear side. *Farm Runner* can carry up to 250 kg (551 lb) with an autonomy of 1.5 hours ploughing and 2 hours cutting and sowing. Cellestial E-Mobility in Hyderabad is working on an electric tractor,apt for horticultural or greenhouse works

The electric agricultural machines (EAMs) described above are only one part of an integrated jigsaw for the future of farming. Several other pieces function in the sky above the farm such as the UAAV (Unmanned Aerial Agricultural Vehicle), from the drone to the satellite. Both have a remarkable history.

Endnotes

[1]Kevin Desmond, Innovators in Battery Technology: Profiles of 95 Influential Electrochemists (Jefferson, NC: McFarland, 2015). Page 237.

[2]Idem, Page 103.

[3]Communication from Stephen Heckeroth, 16th April 2019.

4

Drone History

This chapter traces the history of the electric quadcopter commonly known as the drone, prior to its use in agriculture.

As long ago as 1493, the versatile Italian Leonardo de Vinci, painter, sculptor, machine designer often inspired by his observations of Nature, made a sketch in one of his little notebooks of a screw-like machine next to which he wrote in his mirror-handwriting:

If this instrument made with a screw be well made – that is to say, made of linen of which the pores are stopped up with starch and be turned swiftly, the said screw will make its spiral in the air and it will rise high.[1]

Four centuries later, in 1863, the Viscount Gustave du Ponton d'Amécourt, President of The Society for the Encouragement of Aerial Locomotion by Means of Heavier than Air Machines, published a forty-page monograph in Paris, entitled "The Conquest of the air by propeller. Account of a new system of aviation." In this he took the Greek words *helico* and *pteron*, meaning "spiral" and "wing", and combined them into the word hélicoptère (helicopter). D'Amécourt and his friends were among those very few who were passionately convinced that the future of flying was with *heavier-than-air* machines. To prove their point, having built a fragile flying machine model driven by clockwork springs, they watched it ascend vertically for a few seconds to a height of less than 3 metres (10 feet), only to crash back to the ground.

Among d'Amecourt's circle were two young enthusiasts: a writer called Jules Verne and an engineer called Gustave Trouvé. In the late summer of 1886, Verne serialized his novel "Robur the Conqueror" (French: *Robur-le-Conquérant*), also known as "The Clipper of the Clouds."[2] Robur's *Albatross*has a slender, clipper-shaped hull made of hydraulically compressed paper or cellulose, 30 m (100 feet)long and 3.7 m (12 feet)wide. Above its flat deck stands a veritable forest of slender masts, 37 of them, each with twin, contra-rotating propellers at the top. At the bow and stern were two more propellers. A large rudder steered the "clipper of the clouds" and spring-loaded shock absorbers cushioned its landings. Meanwhile in 1887, Gustave Trouvé demonstrated his 90 gram (3 oz) Lilliputian electric motor which, fitted with an aerial propeller then attached to one end of a scale, once electrified, lifted the scale arm up. Occupying less than a 3 cm (1.2 in) cube, the motor could rise to a height of 22 metres (72 feet) in one second and would enable future experimentation with electric helicopters and aeroplanes.[3]

Still in France, forty years later, the Breguet brothers, Louis Charles and his younger brother Jacques, were developing AC electric motors for submarines at the family business in Douai, France, when they came into contact with the respected physiologist Professor Charles Richet. He challenged them to design a gyroplane with flexible wings. The Bréguet-Richet had an uncovered open steel framework with a seat for the pilot and a 34 kW (46 hp) Antoinette water-cooled piston engine at the centre. Radiating from the central structure were four wire-braced tubular steel arms, each bearing a superimposed pair of four-bladed rotors. In essence, this was a quadcopter. To eliminate the torque effect, two rotor sets were driven clockwise and two anti-clockwise. On 29th September, 1907 *Gyroplane No.I* was given a static test by engineer Maurice Volumard in the family factory and then flown in a field at La Brayelle, albeit to an elevation of only 0.6 metres (2.0 ft) for a couple of minutes! It was not a free flight, as four men were used to steady the structure. It was neither controllable nor steerable, but it was the first time a rotary-wing device had lifted itself and a pilot into the air. It later flew up to 1.52 m (4.99 ft) above the ground. The design was improved and *Gyroplane No.II* appeared the following year. *No.II* had two two-blade rotors of 7.85 m (25.75 ft) diameter and also had fixed wings. Powered by a 41 kW (55 hp) Renault engine, it was reported to have flown successfully more than once in 1908. *No.II* was damaged in a heavy landing and was rebuilt as the *No.IIbis*. It flew at least once in April, 1909 before being destroyed when the company's works were badly damaged in a severe storm.

Another French flight design was inspired by the motion of the maple seed spinning as it falls to the round. In 1913, Alphonse Papin and Didier Rouilly of Paris obtained French and world patents for their "Gyroptère", characterized in the contemporary French journal *La Nature* in 1914 as "un boomerang géant" (a giant boomerang). The machine weighed 500 kg (1,100 lb) including the float on which it was mounted. It had a single hollow blade with an area of 12 m^2 (130ft^2), counterweighted by a fan driven by an 80 hp Le Rhone rotary engine spinning at 1,200 rpm, which produced an output of just over 7 m^3 (250 ft^3) of air per second. The fan also propelled air through the hollow blade, from which it escaped through an L-shaped tube at a speed of 100 m/s (330 ft/s). Directional control was to be achieved by means of a small auxiliary tube through which some of the air was driven and which could be directed in whatever direction the pilot wished. The pilot's position was located at the centre of gravity between the blade and the fan. Testing was delayed due to the outbreak of World War I and did not take place until 31st March, 1915 on Lake Cercey on the Côte-d'Or. Due to the difficulty of balancing the craft, a rotor speed of only 47 rpm was achieved instead of the 60 rpm which had been calculated as necessary for take-off. In addition, the rotary engine used was not powerful enough; it had originally been planned to use a 100 hp car engine, which proved unobtainable.

In 1924, Peugeot electrical engineer Etienne Edmond Oehmichen flew his *Hélicoptère* for 1 km around a closed triangular circuit on the Arbouans flying field, next to Peugeot's auto factory. Its four vertical axis rotors were distributed on either side of the fuselage, while five small propellers ensured horizontal stability, completed by a steering prop and two forward props propelled by an 88 kW. Rhône motor. The Vertical Take Off and Landing, used by today's agri drones born.

Also, in 1915 Croatian electrical engineer, Nikola Tesla described a fleet of unmanned aerial combat vehicles.

Still in France, then the cradle of innovation, Etienne Edmond Oehmichen (as in Ohm) was a keen naturalist studying the flight of birds and insects. He earned his bread and butter as an electrical engineer, first at the Electrical Department of the Mechanical Construction Company of Alsace in Belfort, then with Peugeot at Beaulieu-Valentigney, where he innovated electric lighting and a starting system for automobiles. He also patented an electric stroboscope for inspecting auto parts. In 1920, Oehmichen published a book "Nos maîtres les oiseaux, étude sur le vol animal et la récupération de l'énergiedans les fluides", published by Dunod. ("Our masters, the birds: a study on animal flight and recovering energy in fluids").

In 1921, financed by Peugeot, Oehmichen assembled his first "hélicostat", fitted with two large propellers and a hydrogen balloon to ensure stability and on 18th February, made a flight of 1 minute at a height of 10 metres (33 ft). On 11th November, 1922 he first flew *Hélicoptère Oehmichen No.2*, fitted with twelve propellers (four lifting and eight for direction). He then progressed to four vertical axis rotors distributed on either side of the fuselage, while five small propellers ensured horizontal stability, completed by a steering prop and two forward props propelled by an 88 kW Rhône motor. On 4th May, 1924 Oehmichen made a flight of 1 km (3,280 ft) around a closed triangular circuit on the Arbouans flying field, next to Peugeot's auto factory. The VTOL was born.

The feat enabled him to obtain a grant of 90,000 francs from the Service Technique de l'Aéronautique (STAé) with which he was able to pay back Peugeot. The same year, he made a stationary flight of three minutes, then another with two passengers on board. Oehmichen continued to develop his hélicostats, integrating the technologies of rotary wing and airship, but was unable to win over those responsible for aeronautics. In his conference of 20thMay, 1937 at the French Colonial Institute, Oehmichen stated his firm belief that the hélicostat was the only solution to ensure aerial safety

The word drone, from is used to describe a male bee. Unlike the female worker bee, drones do not have stingers and gather neither nectar nor pollen. The word drone was adapted by British airplane designer and builder Geoffrey de Havilland, also a passionate lover of insects and moths, or lepidopterist. The biplanes which were rolled out of his hangars in Stag Lane Aerodrome, Edgeware, North London carried names such as Hermes *Moth, Genet Moth, Gipsy Moth, Cirrus Moth* and *Giant Moth.*In 1935, de Havilland and his team produced a pilotless version of their *Tiger MothDH.82* which was called the *Queen Bee.* But once airborne, the noise it made reminded "DH" of a drone rather than a queen bee.

In December, 1935 the Chief of US Naval Operations, Admiral William H. Standley, attended the London Disarmament Conference, after which he was given a demonstration of the *Queen Bee*. On his return to the USA, Standley assigned an officer, Lieutenant Cmdr. Delmer S. Fahrney at the Radio Division of the Naval Research Laboratory, to develop a similar system for US Navy gunnery training. In homage to de Havilland, Fahrney adopted the word drone, from then on, the US Navy's official designation for target UAVs in the years to come.

The role of today's drones in aerial spraying or seed-bombarding can also be traced back one hundred years.Although the use of poison gas in warfare had been prohibited by the Hague Conventions of 1899 and 1907, on 22nd April,

1915, less than nine months into the First World War, the German army unleashed a terrifying new weapon that changed the face of warfare forever. At around 5pm, across a 6km (3.7 mi) front, troops released almost 6,000 metal canisters – 168 tonnes (165 ton) – of poisonous chlorine gas towards trenches held by French and Algerian forces near the Belgian city of Ypres. The results were devastating. A noxious yellow cloud enveloped the Allied positions, and within moments 5,000 soldiers were dead, with another 10,000 injured, as the gas ate into their unprotected lungs. By the end of the war in 1918 the allies had used more tonnes of gas than the Germans. In total, chemical weapons killed nearly 100,000 people during the conflict, wounding an estimated one million. The use of chemical and biological weapons was banned after the First World War.

In violation of the Geneva Protocol, Fascist Italy used mustard gas and other "gruesome measures" against Senussi forces in Libya as early as January, 1928 The Italians dropped mustard gas from the air. From October, 1935 to March, 1936 Italy used mustard gas against the Ethiopians during the Second Italo-Abyssinian War, also in violation of the Geneva Protocol. Benito Mussolini personally authorized Italian general Rodolfo Graziani to use chemical weapons at Gorrahei against the forces of Ras Nasibu. Chemical weapons dropped by warplane proved to be very effective as used on a massive scale against civilians and troops, as well as to contaminate fields and water supplies. Some estimate that up to one-third of Ethiopian casualties of the war were caused by chemical weapons.

In peacetime USA, the first known use of a heavier-than-air machine to disperse products occurred on 3rd August, 1921.[4] Crop dusting was developed under the joint efforts of the US Agriculture Department and the US Army Signal Corps' research station at McCook Field in Dayton, Ohio. Under the direction of McCook engineer Etienne Dormoy, a United States Army Air Service Curtiss JN4 *Jenny* piloted by John A. Macready was modified at McCook Field to spread lead arsenate to kill catalpa sphinx caterpillars at a Catalapa farm near Troy, Ohio. The first test was considered highly successful. The first commercial operations were begun in 1924, in Macon, Georgia by Huff-Daland Crop Dusting, which was co-founded by McCook Field test pilot Lt. Harold R. Harris. Use of insecticide and fungicide for crop dusting slowly spread in the Americas and, to a lesser extent, othernations in the 1930s. The name 'crop dusting' originated here, as actual dust was spread across the crops. During the next half century, crop dusting by pilot-flown aircraft and helicopters was used across the agricultural world. Few suspected that this task would later be taken over by pilotless drones.

In 1983, Yamaha began developing pilotless drones at the request of the Japanese Ministry of Agriculture, Forestry and Fisheries and completed its first utility-use unmanned helicopter, the *R-50*. The gasoline-powered aircraft had a two-bladed rotor and was remote-controlled by a line-of-sight user. It was designed primarily for agricultural use, and was capable of precise aerial spraying of crops, going onto treat 970,000 hectares (2.4 million acres) of farmland in Japan each year, particularly in rice paddies.

Elsewhere, isolated farmers such as Robert Blair of Idaho, radio-controlled model aircraft to manage their fields.

The idea of the quadcopter was revived by Richard Vogt, former Nazi engineer and chief designer at Curtiss-Wright's Santa Barbara Division (formerly the Aerophysics Development Corporation). Vogt was asked to design a flying jeep type light VTOL utility vehicle, designated the *VZ-7*. Of exceedingly simple design, essentially consisting of a5m (6ft) rectangular central airframe to which four vertically-mounted propellers were attached in a square pattern. The central fuselage carried the pilot's seat, flight controls, fuel and lubricant tanks, and the craft's single 425shpTurbomecaArtouste IIB shaft turbine engine. In essence a quadcopter, the aircraft performed well during tests, but was not able to meet

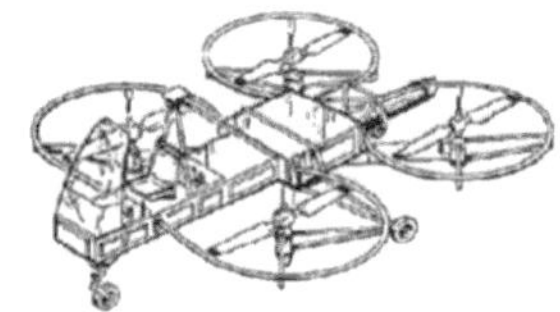

Fig 1. is a perspective view of an aircraft showing my new design

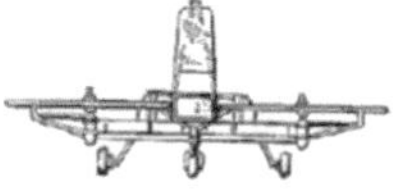

Fig 2. is a front view thereof

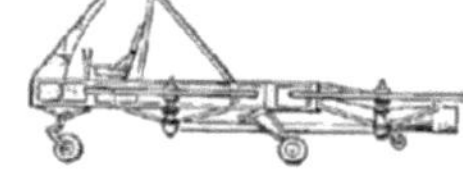

Fig 3. is a side elevational view thereof

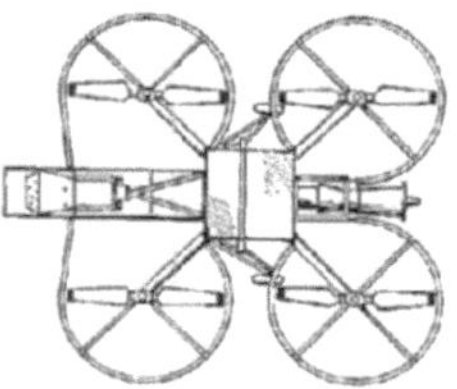

Fig 4. is a plan view thereof

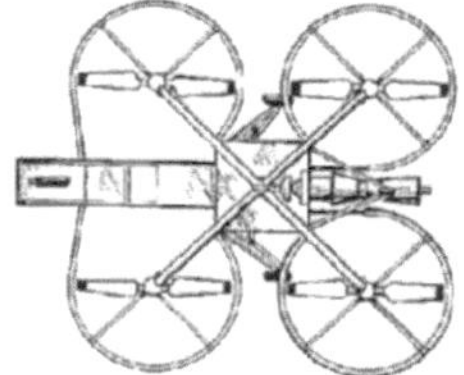

Fig 5. is a bottom view thereof

Fig 6. is a rear view thereof

Richard Vogt's 1960 patent was the ancestor of today's quadcopter variations the Army's standards, therefore it was retired and returned to the manufacturer in 1960. Vogt's Patent US D189462 S, granted in December, 1960 would become the ancestor for today's quadcopter innovations.

With the advent of the energy-dense lithium battery in the early 2000s (as described in Chapter xx), the whole range of transport on land, afloat and in the air, embraced the challenge of electric propulsion. Among them was aero-modelling, including the drone. One of the first companies to build an e-drone was Novadem in France. Since childhood, Pascal Zunino and his friend Fabien Paganucci of Les Pennes-Mirabeaux, in the Bouches-du-Rhône Department of south-east France, enjoyed building electro-mechanical gadgets. Both attending the local la Renardière primary school, in the 1990s they read an article in an aero-modelling magazine about a quadcopter, and decided to build one themselves in the Zunino garage. Moving on to the Jacques Monod College, they continued to work together on projects, with the mechanics developed by Fabien and electricity then electronics by Pascal. At the Pierre Mendès France secondary school in Vitrolles, Zunino and Paganucci assembled a team to compete in the E=M6 robotics competition. They called their robot *Glouton* (Glutton). The circuit boards built were used to liaise between the robot and the control unit through a PWM modulation, and so command the two propulsion engines and that of the suction turbine. Winning, they were officially presented with the 2000 E=M6 Trophy at the City of Science and Industry in Paris. In 2001, while Pascal Zunino studied electronics at the University of Aix in Marseilles, Fabien Paganucci studied micro technology and Computer-Aided Architectural Design in the same city. They assembled the red micro-robot *Z6* prototype drone, which made its very first flight in the early hours of 25th January, 2001. By 2002, they had progressed to their more spindly 70 g (2.5 oz) carbon fibre*X4*. Its electronics brought together on one card an 8051 microcontroller, a 3-gyro inertial unit and 4 speeds. They also built an 80 g (2.8 oz) *Mini X4* with a 34cm (13 in) wingspan and an even smaller 40 g (1.4 oz) *Micro X4* with a 16 cm (6 in) wingspan. This gave them confidence to assemble a team of students at Grenoble INP and to build the rotor-foldable *CPX4* for the international miniature drone competition organized by the ONERA and the DGA that September, 2005. At the Mourmelon Military Camp near Châlons-en-Champagne, they demonstrated the *CPX4* to civil and military professionals in the industry and won first prize ex-aequo. It had particularly attracted the attention of the DGA (Délégation Générale de l'Armement), the French State organization responsible for armament programmes, as being operational and innovative due to its size and previously unseen ease of use: with its hovering capability, the *CPX4* could also be very easily fitted with sensors or cameras (day/night, infrared).

2010, left to right: Fabien Paganucci, Eric Zunino and Pascal Zunino. Pascal Zunino founded Novadem Aerial Robotics in Aix-en-Province with the aim of democratizing drones (NOVADEM).

In 2006, financially backed by the Michel Caucik Innovative Business Incubator, they created the start-up Novadem Aerial Robotics, in Aix-en-Provence with the aim of democratizing aerial robotics. In this they were joined by Pascal's brother, Eric Zunino, himself specializing in Information Technology and who had often helped with previous prototypes. Before long, their enterprise came to the attention of Henri Seydoux of the Parrot Company. In 1994, Seydoux, son of the head of the Pathé-Gaumont Cinema Empire Jérôme Seydoux, together with Christine de Tourvel and Jean-Pierre Talvard had set up Parrot to create a vocal recognition system for the blind in the burgeoning smartphone and tablet market. To diversify their business, Seydoux was looking into a Wi-Fi remote-control automobile. The product was ready for production. But it was not fun enough. Seydoux was about to cancel the project when he got the idea to turn the car into a drone, formed a team called Commando and contacted Zunino and Paganucci at Novadem, who would help Parrot to develop various key aspects of the drone: propeller design, control system and stabilization algorithms, ultrasound telemeter, benchmarks, noise reduction, and testing procedures, while Parrot focused mainly on video. That was the beginning of a time-saving collaboration that would last for three years.

The AR (Augmented Reality) Drone, was officially unveiled at the 2010 Consumer Electronics Show (CES) in Las Vegas as the first Wi-Fi augmented reality quadcopter drone, a form of flying video game. It could be controlled by mobile or tablet operating systems such as the supported iOS or Android within their respective apps or the unofficial software available for Windows Phone, Samsung BADA and Symbian devices. The drone was loaded with sensors and video cameras. It had debuted just in time to surf the wave of the famous, miniature, remote-control helicopter *PicooZ*. In fact, Parrot was not just catching the wave but was extending it. The *AR.Drone* was launched onto the market just a few months later, with a global launch beginning in August, 2010 starting in Hong Kong and France before spreading to the rest of the world. This initial product was a major success: 120,000 *AR.Drone*s would be sold by the end of the year.[5]

Along with *AR.Freeflight*, the application designed for free operation of the drone, Parrot also released *AR.Race*, allowing users take part in solo games, or interact with other drones in combat simulations. Inside the airframe, a range of sensors assist flight, enabling the interface used by pilots to be simpler, and making advanced flight easier. The onboard computer runs a Linux-operating system, and communicates with the pilot through a self-generated Wi-Fi hotspot. The onboard sensors include an ultrasonic altimeter, which was used to provide vertical stabilization up to 6 m (20 ft). The rotors are powered by15 W brushless motors powered by an 11.1volt lithium-polymer battery. This provides approximately 12 minutes of flight time at a speed of 5 m/s (11 mph). Coupled with software on the piloting device, the forward-facing camera allows the drone to build a 3D environment, track objects and drones, and validate shots in augmented reality games.

In 2012, Parrot acquired 57% of the Swiss aerial imaging drone manufacturer of sense Fly4 drones and invested more than 2.4 million Swiss Francs in Pix4D, both in Lausanne. The consumer drone market took off in 2015, but grew slowly and seemed to reach a saturation point. Toy drones had enjoyed a tremendous success but like many fads, at one point it looked like they might go out of fashion unless more commercial uses could be found for them. Media reports said drone makers such as EHang and Zerotech had cut jobs and other costs, and sought to diversify. To counter this, Parrot had invested in diverse drone start-ups. These included Airinov who were developing drones for agriculture.

Airinov founders, Romain Faroux, Florent Mainfroy and Corentin Chéron (Credit Maddyness).

Airinov was the fruit of a meeting in a barn in November, 2009 between Romain Faroux, a farmer's son in the Poitou-Charentes region, withFlorent Mainfroy and Corentin Chéron, drone enthusiasts both from the same engineering school: ESIEA (Ecole Supérieure of Computer Science, Electronics, Automatic). Here, as interns they had been developing open source software to create aerial photographic mosaics using a V-wing UAV they called an *Agridrone*. Could this be used by farmers to observe and analyse the state of plantation? In 2013, Faroux, Mainfroy and Chéron made a proposal and won a prize in the National Competition of Assistance to the Creation of Innovative Enterprises organized by the Ministry of Higher Education and Research and Oséo. Thanks to this prize, they were able to finance a market study, during which they worked with the world-class research institute INRA (L'Institut national de la recherche agronomique) to develop the first-ever multi-spectral sensor to measure crop growth as adapted the solution to farmers' real needs. To quote from CorentinChéron's patent applied for in 2013: "The invention is in the field of image acquisition by a drone to generate a map relating to a state of agricultural crops flown. More specifically, the invention relates to a remotely piloted aerodyne comprising a multispectral imaging device, and a map generation system." From then on Airinov did not look back, Parrot investing 1.6 million euros in the concern in February, 2014. Abandoning the construction of drones to focus on sensor and agronomic models, the company has grown from 3 to 24 employees in four years, is serving more than 3,000 farmers throughout France and has even recently launched its network of operators: the Agridrone network, which has a growing number of members from most of the agricultural world (contractors of agricultural works, technicians of cooperatives or tradesmen, advisers of Chambers of Agriculture). In the drive to make farming ever more precise, Airinov produced their *eBee*, and then the *eBee SQ*, whose one-hour flight capacity is designed to adapt exclusively to Parrot's Sequoia agronomic sensor. Flying over their fields, the drone records four different-coloured "bands" of sunlight that are reflected off the crops at different times of the season enabling farmers to very quickly process the data collected by the Sequoia sensor on the 'first' platform developed specifically by Airinov on its Control Tower. In a few hours maximum, via smartphone, users receive their NDVI cards to observe all their cultures. This, for example, enables the precise control of nitrogen inputs on wheat and rapeseed.

A rival to Parrot-Airinov is DJI (Dajiang Innovation) in Shenzhen, China, directed by Frank Wang. In 2004, Wang's dream of landing at an elite American university such as MIT or Stanford was thwarted due to his unimpressive academic performance, so he ended up at the Hong Kong University of Science & Technology, where he studied electronic engineering. In 2006, during his senior year, working in his dorm room, Wang developed a helicopter flight-control system and even though the hovering function for the on-board computer failed just before the presentation, Li Zexiang, his robotics professor noticed him and brought him into the school's graduate programme. In 2006, Wang and two classmates moved to a manufacturing hub of Shenzhen, Guangdong, China and started working out of a three-bedroom apartment. Wang funded the venture Shenzhen Dajiang Innovation Technology, short for DJI, with what was left of his university scholarship, and modest sales to Chinese universities and state-owned power companies were enough to help him pay for a small staff. In January, 2013 DJI produced their white *Phantom 1* drone. It was a multirotor that most people could order online, read the manual, and fly without much trouble. It was commonly equipped with a GoProcamera for amateur filmmaking or photography. Its battery life was around 15 minutes with a GoPro. Eleven months later, DJI released the *Phantom 2*. Upgrades included auto-return, increased flight speed, increased flight time and controllable range, increased battery capacity, smartphones, tablets to command the drone and a feature called Smart Guidance Control (or IOC), that made the *Phantom 2* ridiculously simple to fly. By the end of 2014, DJI, with a 2,000-strong workforce, had sold an estimated 400,000 units, making Frank Wang the world's first drone billionaire.

The same year, the American Federal Aviation Administration (FAA) proposed some regulations and guidelines to ensure proper operation of these vehicles in the USA, and the regulations came into effect on 21st December, 2015. As the Consumer Electronics Show 2017 was opening, news broke DJI had bought a majority stake in the German camera manufacturer Hasselblad, the company that made the cameras American astronauts took to the moon. Soon after, DJI departed from its white *Phantom* range to present its *H*drone, a black hexacopter with ST16 remote control with Android software and 25-minute autonomy.

In November, 2015 DJI launched its first agricultural drone, the *MG-1*, marking its diversification into the agricultural sector. It unveiled an upgraded agricultural drone, the *MG-1S* in 2016 and the *MG-1S* Advanced in 2017, with upgraded flight control system, radar and sensors. It also provided consumer loans so that buyers could afford drones; its financial services sought to enable operations and related training.

The DJI "Agricultural Wonder Drone" makes it possible for a single farmer to feed or spray pesticide on as much as 32 hectares (80 acres) of crops a day. By comparison, a worker on foot may cover less than 0.5 hectare (one acre) in the same amount of time. Using the DJI drone, the farmer maps out the field by walking through it and the unit's controller creates a flight path.

Bret Chilcott of AgEagle in Neodesha, Kansas with one of his fixed wing drones, an alternative to quadcopters (AgEagle).

Another contender in the market is HoneyComb of Wilsonville, Oregon with its AgDrone System developed by Ryan M. Jensen. Jensen was born in Brazil and given up for adoption as a newborn. His adoptive parents arrived in Brazil intending to adopt another boy, but the baby had died and Ryan, then two weeks old, was next in line. He grew up on a 100-acre farm outside Estacada. He knows nothing about his biological parents, but displayed an early aptitude for maths, science, rockets and computers. At fourteen he skipped high school and enrolled in community college, becoming the first in his adoptive family to attend college. By nineteen, he received a BSc. in Mechanical Engineering and minor in Electrical Engineering. Before his Masters' Degree, he worked with NASA and astronauts aboard the International Space Station. In 2012, Jensen teamed up with Ben Howard and John Faus to create HoneyComb. Their flagship product, the V-wing AgDrone System, is powered by a 575W electric motor using 8000m/Ah Li-Po battery for up to 344 ha/hr(858 acres/hr) at an altitude of 122 m (400 ft). Six-channel image processing, one-click data handling, robust algorithms, and optimized hardware configuration provide a seamless, durable, and easy-to-use system for drone-based mapping and scouting.

The fixed wing UAV with its pusher propellers has been used to advantage by Bret Chilcott at AgEagle in Neodesha, Kansas. Growing up on a farm with a love of playing with radio-controlled model aircraft, while Chilcott worked first for Cessna Aircraft, then for Cobalt Boats, and also head of R&D at Snap-on, he learned the advantages of carbon fibre-infused construction using vacuum-assisted resin transfer. In 2010, Kansas State University began collaboration with Chilcott to merge small radio-controlled aircraft and near infrared photo technology to determine crop health. In 2011, Chilcott founded AgEagle and travelled across the corn belt of the US demonstrate his aluminium catapult-launched, white and yellow electric fixed wing drone, whose carbon fibre construction render it tractor tough, with a flight time of about forty minutes. From this prototype, Chilcott and his team developed the 3 kg (7 lb) light *RX60*, able to capture virtually thousands of high resolution NIR/NDVI aerial images and cover up to 162 hectares (400 acres) on a single battery charge in the service of precision agriculture growers and agronomists. Flying by itself, all flight plans can be quickly set up anywhere at any time, using only an Android phone or tablet; and with a cellular data subscription, users are able to check the results within minutes of flight and perform crop scouting while they are still in the field. The *RX60* can capture the images even during the cloudy weather. In 2016, AgEagle produced its lighter *RX47* and *RX48* drones. The three models have been sold across North America as well as Australia, the U.K. and various other countries through independent precision agriculture experts. In 2018, they unveiled drone-enabled data collection and analysis solutions to support food manufacturers' sustainability solutions with reduced chemicals. AgEagle also acquired Agribotix including its FarmLens data analytic platform.

In April, 2019 AeroVironment of Monravia, California, who had been building SUAVs since the 1980s and its military *Ravens* and *Pumas* since 2000, announced the agricultural version of its fixed-wing/multirotor *Quantix* hybrid drone and AeroVironment Decision Support System (AV DSS). This offers farmers the first fully integrated drone, sensor and software information solution, as easy to use as an app, for collecting and using aerial imagery. AeroVironment launched the product at three tradeshows: the World Ag Expo, National Farm Machinery Show and Commodity Classic. *Quantix* can fly up to 162 hectares (400 acres) per 45 minute flight, with a cruising speed of 64 km/h (40 mph).

AeroVironment's Quantix, is the direct descendant in a lineage of Small Unmanned Aerial Vehicles which can trace their origins back to the 1980s (AeroVironment).

Drones flying solo across fields was already smart, but what if they could fly in swarms? Lorenz Meier, a PhD student in electronics at the ZTH Swiss Federal Institute of TechnologyZurich was curious about technologies that could allow robots to move around on their own, but in 2008, when he started looking, he was unimpressed. Most systems had not yet even adopted the affordable motion sensors found in smartphones. So as part of what he called the "sFly Project: Swarm of Micro Flying Robots", Meier built his own system instead: the Pixhawk flight stack platform for drones for autonomous drone control. In 2009 he launched,its results inspiring several drone companies such as Parrot, DJI, Yuneec, Skydio. Importantly, Meier's system aimed to use cheap cameras and computer logic to let drones fly themselves around obstacles, determine their optimal paths, and control their overall flight with little or no user input. In 2011, he open-sourced it under the PX4 software brand following through with the Pixhawk open hardware autopilot in 2013. The following year, the Dronecode Opensource UAV Platform, which is a non-profit organization governed by the Linux Foundation was formed by Chris Anderson with the goal of using open-source Linux for the benefit of founding members including Qualcomm and Intel. 3D Robotics moved along with Yuneec International providing cheaper, better, and more reliable UAV software. For example, the Yuneec DataPilot™ software, integrated in the new *H520* hexacopter, is tightly coupled to the PX4 flight control architecture. The DataPilot™ is a complete solution for planning survey and waypoint-based UAV flight.More than 500,000 units of Pixhawk have been manufactured. In February, 2017 Meier helped Yuneec to establish its R&D (research and development) centre in Zurich and since then has become an advisor to the company in high-tech development. In September, 2017 Meier was named as one of top 35 innovators by *MIT Technology Review*.

During the past four years, the applications of the electric agricultural airborne machine (EAAM) have become so diverse that it merits an entire chapter of its own.

Endnotes

[1]Leonardo da Vinci, Codex Atlanticus B., fol 83 verso.

[2]Jules Verne "Robur Le Conquerant" *Journal des débatspolitiques et littéraires*, June 29 to August 18 1886, then in book form (Paris: Hetzel et cie) 1886.

[3]Desmond, *Trouvé*, 111.

[4]Mary AnnJohnson,. *McCook Field 1917-1927*. Landfall Press, Dayton, Ohio: 2002

[5]Communications from Novadem during July 2016.

5

Drones: Today and Tomorrow

EAAMs (Electric Agricultural Aerial Machines) have emerged from drones used for both aerial filming, as toys and even as racing units. A Price water house Coopers' study revealed that the global market for drone-powered business solutions was valued at $127.3 billion in 2016. For agriculture, prospective drone applications in global projects were valued at $32.4 billion. Stelios Kotakis and colleagues at market research firm IHS Markit projected that there would be around 400,000 shipments of drones to firms in the agriculture and forestry sectors in 2017. The Food and Agriculture Organization (FAO) of the United Nations projects that by 2050 humanity's ranks will likely have grown to nearly 10 billion people. Farmers will need to produce more with less, while preserving our environment for future generations. And society has a duty to help them achieve this. Although agriculture is perceived as a traditional economic sector, precision agriculture technologies have already boosted crop yields significantly in the last decades. The same study forecast that agricultural consumption would increase by 69% from 2010 to 2050.

Today's multi-purpose agri-drones fly in formation (DJI).

Droneshave vital part to play in this challenge, with their increasingly wide variety of applications, from observation for soil and field data analysis, to tree plantation, pest control, crop monitoring or scouting, to cattle monitoring, to fence inspection, to checking a centre pivot or linear irrigation, to fertilisation, irrigation, to pollination, to transportation and health assessment. This chapter outlines their range.

Observation for soil and field data analysis

In 2019, Israel Aerospace Industries Ltd. (IAI) signed an agreement with Brazilian company Santos Lab on use of drones (UAVs) and advanced analytics for large scale precision agricultural applications. Under the terms of the agreement, IAI will provide the UAV systems and analyse the collected data. Santos Lab will operate the *BirdEye 650D* UAV, including its hyper-spectral wide coverage imager, a unique development by IAI for the precision agriculture market. The UAV will be used for generating reports on these large-scale farming areas according to a broad range of parameters. The *BirdEye 650D* will perform agricultural missions covering large areas utilizing beyond visual line of sight (BVLOS) missions, with typical tasks including monitoring various crops such as soy and sugar cane, as well as commercial forestry. The reports will include accurate analysis of the crops and soil condition. The data will be available for customers through a dedicated cloud solution, including high precision analytics that would not have been possible without the hyper-spectral technology, which allows identifying crops' condition from high altitude at high resolution and accuracy. The service is expected to become operational at the end of 2019. Since August 2018, in the small Central American nation of El Salvador, many fields, mainly of sugar cane, are now being tended by *Hylio Agrodrones* using AgroSol command software.

Emile Faye, a French researcher in digital agro-ecology and a team at CIRAD, the French Agricultural Research Centre for International Development have teamed up with the Senegalese Institute for Agricultural Research to help farmers who have until now estimated their mango crop by counting the fruit on a bunch of trees and then extrapolating for the whole plantation. This rough-and-ready method has considerable room for error. HortSys's *Pix Fruit* alternative uses advanced modelling software to produce a more precise count of the crop. Using a smartphone, the farmer takes photos of a selection of trees in his fields. Fruit-recognition technology then calculates the likely overall harvest, drawing on a tree canopy height and land cover databank compiled with the help of drones that also includes information on climate, soil and administrative constraints. A methodological toolbox was developed and tested to estimate and map tree species, structure, and yields in mango orchards of various cropping systems (from monocultivar to plurispecific orchards) in the Niayes region of West Senegal. The system could be extended to coffee, lychees and citrus fruits.

Forty-five students along with five faculty members from the Electronics, Mechanical and Agricultural Engineering department, Lovely Professional University (LPU), Punjab, India have conceptualized and designed ***Flying Farmer***, a drone that can be exclusively deployed in mapping and survey of yields and biomass. It also estimates the nutrient content of the soil to aid production growth and lessen crop damage. LPU will not file for a patent, but instead will open source the technology so that it can be available to any farmer, anywhere at a reasonable price.

During 2019, scientists from the Norwegian Defence Research Establishment (FFI) and Paul Hellhake at Rajant Corporation Malvern, Pennsylvania, well known for its wireless Kinetic Mesh® networks, have been working on simultaneously flying about 20 drones that can work in co-ordination with little human supervision for up to 30 minutes or less. Once tractors, fleets, UGVs, soil sensors, and more are connected to Rajant's peer-to-peer network via BreadCrumb nodes, they form an autonomous Industrial Internet of Things (IIoT) environment. If a farmer needs to map hundreds of acres of their field, it might take 50 battery charges. A swarm is essentially doing in parallel the same job up to 20 times faster than a single drone.This "parallelisation of the workload" will single out healthy plants from sick ones and help farmers decide where to deliver more pesticides and nutrients.

Tree Plantation

Every year, 15 billion trees are destroyed from natural and anthropogenic causes. Despite US$50 billion a year spent on replanting, there remains an annual net loss of 6 billion trees. Governments have made commitments to restore 350 million hectares of degraded landby 2030, equivalent to an area the size of India, which could accommodate around 300 billion trees.As rates of deforestation and land degradation continue to rise, a 2018 report from the World Resources Institute and the Nature Conservancy identified ecosystem restoration as a growing area of economic opportunity for businesses, and attractive for investors as well.

Start-ups are experimenting with using drones to protect the Amazon rainforest, by employing tools to collect data information from hard-to-reach places. Drones can gather an enormous amount of data from the rainforest, and a non-profit called the Amazon Conservation Society is using them to help manage a large swathe of rainforest. Researchers use AI software to compare two different images of the forest, where the overlay can show discrepancies and alert researchers to whether logging is taking place. It doesnot always mean logging is happening it could be farm clearing or another activity, but it helps researchers focus their attention on a smaller space in what is an extraordinarily large area.

Start-ups have created drone-planting systems that achieve an uptake rate of 75 percent and decrease planting costs by 85 percent. These systems shoot pods with seeds and plant nutrients into the soil, providing the plant all the nutrients necessary to sustain life. Two companies are using drones to step up the rate of tree-planting are BioCarbon Engineering founded by Lauren Fletcher and DroneSeed, founded by Grant Canary.

During the late 1990s, Lauren E. Fletcher, with a Master's Degree in Civil and Environmental Engineering was a space-systems engineer at NASA Ames Research Center, specialising in bio engineering. In 2007, he was at the International Space University, then from 2008 to 2010 at Stanford University. From 2010 to 2019 Fletcher was a Doctoral Student at Oxford University's department of Physics on "Project Mars on Earth". In 2009, while Fletcher was at COP15 in Copenhagen, he became more concerned about the state of our world: degrading climate, loss of natural environments, significant biodiversity losses, and a potential for global scale human suffering. After years of studying climate change and the environment, Fletcher asked himself how the damage of more than a century of anthropogenic development could be reversed. The answer, in part, is restoring the planet's decimated forests, to counter industrial scale deforestationusing industrial scale reforestation. In 2013, Fletcher linked up with businessperson Susan Graham with a PhD in healthcare innovation to foundthe company called BioCarbon Engineering (BCE), based in Eynsham, Oxfordshire, UK to plant at least 1billion trees a year with drone swarms. To do this needed a technician.

Enter French drone engineer, Jeremie Leonard. From 2005 to 2007 Leonard studied at the Lycée Marcelin Berthelot, Saint Maur des Fossés, France, then at the Ecole Supérieure d'Electricité, at Gif Sur Yvette, Isle de France. He then crossed the English Channel to study for his PhD at Cranfield University, between 2011 and 2014, where the aim of his thesis, named "Project Athena", was to develop a fully autonomous swarm of medium-altitude long-endurance Unmanned Aerial Vehicles (MALE UAV) with integrated health management. Leonard's work encompassed research on mission planning, multi-agent control and swarm energy management. In 2014, Leonard was recruited by Fletcher to BioCarbon Engineering.As Leonard explained:

> I have been fascinated by the advances in the drone industry. It started as a few university research programmes, gained popularity through some toy-level products and is now being used across a large number of fields and revolutionise them. Specifically, the development of remote sensing and automated systems in forest restoration excites me most. This is also why I am doing my job as an engineer and a drone pilot of ecosystem mapping and tree-planting drones.

The "seed-dropping" system developed by BCE uses satellite and drone-collected data to determine the best location to plant each tree. The planting drones fire a biodegradable seedpod into the ground with pressurized air at each predetermined position at 120 seedpods per minute. They fly at an altitude of 1 to 2 metres (3 to 6.5 ft) above the ground. A small pressurized canister provides the necessary propulsive force for the seedpods to easily penetrate the soil's surface. The seedpods are filled with a germinated seed, nutritious hydrogel and other vital components. The pods break open upon impact allowing the germinated seeds to grow. These penetrate the earth and, activated by moisture, grow into healthy trees. Two operators equipped with 10 drones can plant 400,000 trees per day. Just 400 teams could plant 10 billion trees each year, with the capability to scale to tens of billions of trees annually. The fully automated and highly scalable BCE solution plants 150 times faster and 4-10 times cheaper than current methods. This technology provides a new tool enabling global enterprises and governments to meet their restoration commitments. BioCarbon Engineering's technology also goes beyond trees to plant other flora native to an ecosystem.

With initial funding in 2016, a patent "for automated planting" was applied for by Fletcher and his team. BCE began its full commercial operations with the first paid project in May, 2017 at abandoned mine sites inDungog in the Hunter Valley, New South Wales,Australia that were in need ofreforestation. They have executed nine projects in the UK, Australia, Myanmar, New Zealand, South Africa and Morocco.

Environmentalists in Myanmar used to plant mangroves by hand. Myanmar has lost at least 1 million hectares of mangrove forest over the past several decades, making it more vulnerable to cyclones and climate change. Since 2012, Worldview has been able to plant over six million trees, which is a huge achievement already. However, with the help of the BCE drones, they could plant another four million by the end of 2019. Since the drones began their work in September, the saplings have grown to be 50 cm (20 in) tall.

In April, 2018 BCE received a funding boost of US$2.5 million. The seed investment comes from SYSTEMIQ, a purpose-driven investment and advisory firm that aims to tackle economic system failures, and Parrot, the leading European drone group. In May, 2018 Jeremie Leonard travelled to Canada to work with the Canadian Forest Service for the first-ever Canadian trial of using drones to plant tree seeds in northern Alberta. Work in 2018 will expand to projects in the UAE, Canada, USA, Brazil, Peru and Spain. Customers include private landholders, companies, nongovernmental organisations and governments. BCE plans to plant to plant 500 billion trees by 2060.

BCE are not alone. DroneSeed based in Seattle, Washington, alsocommitted to reforestation efforts, has developed a plan for each planting area that maximises successful planting and tree growth. Understanding the environmental conditions of the site is paramount to successfully replanting the area. Using Lidar, topographical 3D maps are made, photographs are taken with a multispectral camerato collect visual data, much of it outside of the realm of human detection, which can then be used for an analysis of the plants and soil before any planting can take place. Using this data, actual planting locations are determined so that each seed package has a much greater chance of survival. With the resulting map, the drones fly autonomously, as many as five at a time, and are supported by a team that is ready to load up the drones and is there in case of any setbacks. The drones use machine learning models, setting out to find various 'microsites' where the seeds will face better chances of survival. The seeds are pre-packaged into small bundles, filled with nutrients and covered in the chemical capsaicin to keep hungry creatures at bay. It is this extra attention to detail which improves the odds of each tree's future success.

After planting, the location is monitored and growth is optimized with fertilizer, herbicide and water, all of which are also applied by the drones. In addition to gathering data needed for planting, drones are also collecting data on growth, canopy cover and other factors which allow the creation of 3D models of the actual reforested area.

Drone Seed founder, Grant Canary M.A.of Seattle, Washington is an environmentalist with a love of outdoor sports. He has spent his entire carrier working within for-profit companies to benefit the environment including Vestas Wind Energy and the US Green Building Council. He raised $10M and built a 5,600m^2 (60,000 ft^2) factory to pioneer the commercialization of black soldier flies (Hermetiaillucens) to treat food waste and produce a sustainable supply of nutrients for sustainable salmon feed and agricultural uses. He also founded BioSystems LLC, a wholly owned subsidiary of Enterra, based in Portland, Oregon. At a loss for what to do next in his career and was told by a friend that perhaps he should just go and plant trees.

Realising that tree reforestation needed intensifying, Canary founded DroneSeed. He recruited Matthew M. Aghai as his Director of Biological Research; John Thomson, adrone systems engineer, responsible for specifying, designing and manufacturing heavy-lift flight systems and supporting hardware to enable company operations; and Robert A Krob, a software engineer. They were soon joined by Matt Kunimoto, a drone systems technician who had built a hexacopter drone that uses image recognition to guide its flight autonomously in order to follow a custom pattern.

In 2015, DroneSeed first won the Beaverton, Oregon $100K Challenge sponsored by the City of Beaverton and Oregon Technology and Business Center. Shortly after, they were one of the nine start-ups selected for Techstars Seattle 2016 out of over 1,000 applicants to the program.With funding from Techstars, Social Capital, and Spero Ventures, to the tune of $4.8 million, DroneSeed received the FAA's first approval for up to five aircraft to be flown by a single pilot each carrying a 57lb payload. The FAA classifies this exception as "precedent setting", referring to the exceptional lengths DroneSeed has gone to prove out its ability to scale operations to larger payloads for multiple concurrent flights. At the time, no other drone operator in the USA could legally operate with such heavy lift aircraft.

The firm works for 3 of the 5 largest timber companies and recently signed a contract with The Nature Conservancy to restore post wildfire burn sites to combat the spread of wildfires and keep affected areas healthy. Their first planting project was in October 2018, replanting after the Grave Creek Fire which burned 7,000 acres near Medford, Oregon in 2018.

In 2018, the DroneSeed team was granted Patent N° 10,212,876 for "Aerial deployment planting methods and systems for making good use of recently obtained biometric data and for configuring propagule capsules for deployment via an unmanned vehicle so that each has an improved chance of survival."

In 2019, DroneSeed and the Nature Conservancy Oregon geared up to save the ecosystem across the Pacific Northwest from invasive species. Drone swarms of up to five aircraft will be deployed to restore rangelands by re-seeding threatened areas – especially in sagebrush steppe habitats. Invasive weed species harm the sagebrush steppe, resulting in a huge swathe of plant loss. In fact, only 50 percent of such plants still exists, with the remaining 50 percent at risk of being lost in just the next 50 years.

Pest control

Invasion by alien species is a worldwide phenomenon with negative consequences at both natural and production areas. Pest damage is one of the most significant problems facing farmers. It is estimated that as much as 35% of yield loss is due to pre-harvest pest activity. Globally, farmers spend over $40 billion per year on pesticides - and the number in the US is almost 25% of that number at $9.2 billion. Why are farmers buying so much pesticide? They are trying to avoid an estimated total of $200 billion in crop loss annually caused by pests. The primary users of the Pest Management Guidelines are Pest Control Advisors (PCA). PCAs are certified experts that help identify pest damage, determine the cause, and recommend a remediation path to fix the problem. The PCA and the farmer work together to get pest problems identified and remediated in the best way possible to minimize pest damage. Pest problems that are not identified and remediated can go from minor problems to major problems very quickly. This makes identifying the initial pest outbreak in a timely fashion crucial to farmers and PCAs – it is arguably the single biggest factor in controlling the amount of pest damage.

This is where pest control drone technology can help. The recent emergence of drones has led to a variety of new solutions for drone pest control management. In many cases, drones offer a solution that is both more effective and significantly cheaper than any current options.

Farmers are using infra-red camera carrying drones to pinpoint problem spots with insects and aphids in vast fields and ranchlands. Based on the mapping, another drone then drops a 'cocktail' of predatory insects, transported in a sock attached the underbelly of the drone and containing a mixture of vermiculite and insects onto grape vines and citrus trees to combat pests. By focalizing pest control, they prevent spread and save money.

After a successful joint venture, in January of 2018 SkySquirrel Technologies and VineView Scientific Aerial Imaging Inc. merged to form VineView. VineView drones can check 50 acres of vineyards in 24 minutes for telltale signs of mold, bacteria or other diseases. The system is used in two of the world's top wine regions - California and France.

XAG, founded by Peng Bin, accounts for more than half of agricultural drone sales in China, serving more than 1.2 million farmers. XAG has 27,000 drones in operation, most of which were sold to companies that offer pesticide-spraying services. For example, 3,000 XAG crop dusting drones are in use the cotton rich Xinjiang province in China. At least 1,797 XAG agricultural drones were used to spray cotton defoliant on about 446,667 hectares of cotton fields in Xinjiang in 2018, which accounted for 30 percent of the total area of machine-harvested cotton fields in the region.

Peng Bin of Guangzhu, China beside one of his XAG drones, accounting for more than half of agricultural drone sales in China, serving more than 1.2 million farmers.(XAircraft).

Peng's parents moved from the countryside to Sanming City in China's southern Fujian province early on. Peng himself was born and raised in the city. A computer science graduate from XiDian University in Xian, he worked for Microsoft for two years before starting XAircraft in Guangzhou in 2007 to make consumer drones. In 2010, XAircraft released the "milestone product" X650 Quad-copter, accounted for 70% of the multi-rotor global market. In 2014, XAircraft founded XPLANET service company, stared to provide plant protection services to famers. By 2017 XAircraft provided plant protection and land surveying service to more than 230,000 farmers. Peng renamed his company XAG in 2018. The Chinese government has set a target to produce 90% of its own farming equipment by 2020, and the sector is one of the priorities in Beijing's "Made in China 2025" strategic industrial plan. To stay ahead of the competition, Peng is looking beyond China for markets for his agricultural drones. XAG has more than 20 Research Stations globally, including Japan, Korea, Thailand, Indonesia, Myanmar, Tajikistan, Kazakstan, Australia, New Zealand, United States, Brazil, Ecuador, Peru, United Kingdom, Germany, Ghana, Nigeria, Morocco, Zambia, etc. Before officially entering each market, XAG's products will be field-tested for not less than 3 farming seasons locally, and be commercialised after access allowance granted and complied with local regulations. Just one example, Zimbabwe in Africa, a huge exporter of tobacco and other crops in Africa's SADC region, is maximizing its productivity by integrating the use of drones for agriculture in its various farming activities. XAG's Global Research Partners include China Agricultural University, Harper Adams University, The University of Sydney, South China Agricultural University, The University of Tokyo, The University of Queensland, Canada A&L Laboratory, Bayer Crop Science Research Centre, Corteva, Australian Centre of Agricultural Research, Tel Aviv University, Wageningen University, Saint-Gobain Materials Science Laboratory and many other research teams in agriculture science and technology. In 2019, to promote their XAG brand, Peng flew to Los Angeles with co-founder Justin Gong, rented an Infiniti QX80 SUV, and went from the wine country in California's Napa Valley through Nevada, Utah, and Wyoming. There are currently 8,500+ XAG drones in operation around the world.

In Japan, almost two-thirds (65.2%) of farmers are over 65 years old. Japan's rapidly aging society and the steady migration of young citizens to the cities pose a significant threat to the country's agricultural traditions and the industry's survival. To address the problem farmers are turning to "smart agriculture" including drones. The Agri Drone developed by Saga University with Saga Prefecture located on the island of Kyushu, Japan, and IT firm OPTiM in Minato, Tokyo. Agri Drone uses a suspended bug zapper which delivers a glowing, insect-enticing electric payload to the points at which the pests are congregating in harmful numbers. Primarily used at night, it utilizes infrared and thermal cameras to shoot targeted doses of pesticides where insects are congregating. OPTiM is working with local governments to help educate about the benefits of drone usage, and is working with Fujieda City in Shizuoka and Obihiro City in Hokkaido Prefecture to find farmers to train.

Optim Cloud IoT company has divided into air monitoring with management (Agri Manager) and a crop record assistant (Agri Assistant). Through aerial photography and sensors, Agri Manager can remotely confirm temperature, sunshine volume and other data. Using image processing technology, it can discover and predict 27 different pests and diseases. The Agri Assistant uses voice recognition technology to upload comprehensive crop information that will reduce farmers' workloads and increase planting.

Drone Volt based in Villepinte, France. known for designing drones for specific uses, like shooting 360-degree video for virtual reality, has introduced the Drone Spray Hornet to locate and destroy the nests of Asian hornets that are becoming a nuisance in parts of Europe. The Drone Volt quadcopter weighs about 3 kg (7 lb), has a payload capacity of up to a 750 ML aerosol can of insecticide, emergency parachute in the case of failure, and battery life of 9 to 18 minutes depending on the number of batteries onboard. Not that the do not fight back; a DJI drone inspecting a suspected nest on the Channel Island of Jersey came under attack as pest control tried to get a closer look. It's thought that some 6,000 hornets were drawn out by the sound of propellers and almost downed the £4,000 aircraft by swarming its propellers and spraying it with their venom.

In 2018 China, DJI doubled its domestic sales of its agricultural drones, the Mavic 2 and Agras MG-1, last year, resulting in 20,000 unmanned aircraft sold to spray pesticides in China, and about 2,000 in Japan and South Korea. In February 2018, CortevaAgriscience, the Agriculture Division of DowDuPont, announced that they would be deploying the world's largest drone fleet with more than 400 DJI drones, using DroneDeploy's software, "Live Map" giving real-time insights to plant health, will change the way farmers manage their crops in the U.S., Europe, Canada, and Brazil.. To put this in perspective, the company explains that in less than 15 minutes a 160-acre field can be surveyed by drone, quickly spotting variations in plant and soil health. All drone operators will be trained to get the most out of the unmanned aircraft and will be certified according to local aviation regulations.[1] As one example, a group of farmers in Shangqiu village, Henan Province China recently bought thirty DJI's MG-1 drones hoping to start a pest-killing business. The drone follows a pre-programmed flying route and is equipped with three sensors Controlled via a wireless console, the aircraft can spray up to 10 litres of pesticide through four nozzles.

In Hawaii, crews of the Big Island Invasive Species Committee (BIISC) have endeavoured to control silk oak trees on the slopes of Pu»uWa»awa»a (literally translated to "many furrowed") near Kona. Using a drone, the crews can find the best path across the landscape, saving time and allowing them to more efficiently do what they do best: get rid of invasive plants. BIISC relies on UAVs to help survey for rapid »ôhi»a death, a fungal disease affecting »ôhi»a trees on Hawaii Island. The main symptom: the leaves turn brown almost overnight, as though frozen in place.

To detect invasive aquatic plants, in 2015, Tao Tang, professor of geography and planning, employed drones to search for invasive water chestnuts in Tonawanda Creek and the Erie Canal. The project, funded by the Great Lakes Research Consortium, is designed as a "proof of concept" investigation. Tang operated the drone from boats from the Great Lakes Center Field Station to take images of 100 sections of shoreline, each 500 meters long.

Pests can also take the form of invasive vertebrates or wild animals, feral dogs, pigs, possums, deer and rabbits which play havoc with crops. To farmers these invasive pests that cause billions of dollars of damage to farmland and animals like sheep and cattle. With drone equipment supplied by Bluebird Aerosystems from Israel, Ninox Robotics of New South Wales, Australia, and Aeronavics of Raglan, New Zealand are tackling this problem. Island Conservation on the Galapagos Islands has been launching drones and hoppers with applicators designed in 3D printers, dropping rat poison hidden within special bait. The inaugural missions have covered 52 percent of North Seymour Island – making it the world's first use of drones to remove invasive vertebrates.

In late October 2018, Beijing TT Aviation, a specialized UAV industry chain service and UAV application solutions provider, participated in forestry pest control for dealing with webworm, microelalopha, and moth, in three cities of Shandong Province, China.

These are just a few examples.

Health assessment

It is essential to assess crop health and spot bacterial or fungal infections on trees. By scanning a crop using both visible and near-infrared light, drone-carried devices can identify which plants reflect different amounts of green light and NIR light. This information can produce multispectral images that track changes in plants and indicate their health. A speedy response can save an entire orchard. In addition, as soon as a sickness is discovered, farmers can apply and monitor remedies more precisely. These two possibilities increase a plant's ability to overcome disease. And in the case of crop failure, the farmer will be able to document losses more efficiently for insurance claims.

In 2015, research engineers Charles Nespoulos and Cyril de Chassey, having worked together at Airbus, founded the Chouette start-up to promote their *GreenSeeker* based autonomous quadcopter for analysing the condition of the vines and grapes in vineyards including mildew, Esca and golden flavescence. Although they started with lesser known vineyards, in July, 2018 Anthony Appolot the technical director oenologist for the Pape-Clément and Lussacvinyards, one of the most prestigious and oldest vineyards in Bordeaux which made its first harvest in the year 1252, began to use the Chouette system on 70 hectares (173 acres) of the vineyard of Bernard Magrez.

AgroHelper, founded by Bulgarian engineer Mihail Marinov, has developed a web-based solution that helps farmers process drone captured images and detect in real time, zones with potential crop health issues. The platform is powered by a Cloud infrastructure and does not require any specific hardware to be present on the farmer's local machine (or a fast internet connection). Currently, most state-of-the-art software solutions use a process called "stitching" to create an orthophoto map from hundreds of individual overlapping aerial photos. Each individual photo captured by the drone camera contains different terrain features such as crop rows, tractor trails or buildings. As the photos overlap, each separate feature is captured by the drone camera multiple times from different angles and perspectives. AgroHelper's Health Map feature indicates zones with potential issues. As the process is optimized, the cost is contained, and AgroHelper provides farmers the opportunity to process three maps per month for free.

Fish Farming

Transportation

In March, 2018 in East China's Zhejiang Province, a swarm of drones belonging to Alibaba's Cainiao ET logistics laboratory transported freshly-picked Longjing tea leaves from Shifeng Mountain, next to Hangzhou's West Lake, to the tea frying centre in less than two minutes. At the centre the leaves are spread out in the sun and dried. Dating back more than 1,200 years, tea pickers had normally spent an hour climbing over the mountain. The potential of using small and medium-size drones to transport harvested crops over difficult terrain remains to be exploited.

Fruit picking

Founded in 2016, Israeli start-upTevelAerobotics Technologies is a developer of autonomous drones equipped with a one-metre (3 ft) long mechanical claw that can pick fruits, or be used for thinning and pruning tasks in orchards. The company currently employs 15 people. Tevel's drones are equipped with artificial intelligence capabilities in order to be able to detect fruit type, blemishes and the fruit's quality based on its ripeness. While the company's current version is able to pick only apples and oranges, Tevel is working on expanding its drones' capabilities to include additional fruits such as avocados and mangoes. The company is currently working on a patented fleet of drones that will be commercially available in 2020.

Crop monitoring

Adama Agricultural Solutions, an Israel-based manufacturer of crop-protecting chemicals, has joined forces with drone maker Tactical Robotics Ltd., based in Yavne, Israel to study the feasibility of developing an unmanned vehicle for the aerial spraying of crops. Tactical Robotics's *Cormorant* drone, formerly known as the *AirMule*, can perform a number of tasks, from putting out fires to rescue missions in tight urban environments.

The drone is a compact and autonomous vehicle that can carry loads of up to 500 kilograms (1,100 pounds), taking off vertically like a helicopter. Because it utilizes internal lift rotors that enable it to operate in obstructed terrain, like mountainous, wooded and urban areas, the drone can fly where helicopters are unable to operate. Under the joint venture, Adama and Tactical Robotics will work together to develop the *Ag-Cormorant*, an unmanned vehicle that will allow aerial spraying of crops both night and day with low levels of noise.

The *Cormorant*'s ability to adjust its speed and altitude to conditions on ground and the fact that the rotors are at the bottom of the vehicle will enable better dispersion and penetration of the chemicals on the crops, according to the company the statement.(Tactical Robotics is a wholly owned subsidiary of Israel's Urban Aeronautics Ltd., a maker of vertical take-off and landing aircraft.)

Formerly called Makhteshim Agan, Adama is now a subsidiary of China's ChemChina. It is a maker of herbicides, insecticides, fungicides, plant growth regulators and seed treatments. Adama had sales of $3.9 billion in 2018

A tech start-up called WeFlyAgri is using drones to map cocoa plantations and track crop diseases in Africa's Ivory Coast. Since an official launch in July, 2018 WeFlyAgri, has started working with a 3,000-member cocoa cooperative to map more than 9,000 hectares (22,000 acres) of land outside Abidjan. The work could form a blueprint for other farmers in the industry and perhaps even help increase access to loans if farmers can provide the mapped evidence of their plantations as collateral.

Cattle monitoring

The use of drones for monitoring livestock is slowly gaining momentum in various countries. Australia and Israel have already started using lots of cattle monitoring drones. Farms, where drones are keeping an eye on the livestock, are in less need of manpower; instead, they can easily keep track of their animals without getting into their trucks or hiring labour on horseback. Each cow or sheep can be fitted with thermal sensing electronic identification earring which can be selected up to 12 m (40 ft) away by the drone, and which can also be fitted with an infra-red camera enabling it to see which animal is feverish by a rise in its body temperature. The drones provide clear thermal images which easily reveal the difference between one animal vs. another.

Scientists within the Texas A&M AgriLife Research University System are testing new technologies at a feedlot in the Texas Panhandle to find ways to reduce the use of antibiotics in livestock and provide consumers with a healthy meat supply. Drones equipped with thermal imaging cameras have been buzzing over a research feedlot near Amarillo, as researchers develop test methods to identify feverish animals before they show symptoms of illness, such as eating less feed or infecting other animals. Engineers from Texas A&M believe that producers will be able to detect sick animals earlier and target their use of antibiotics more precisely than is possible with current technology.

Ranchers have also found that by adding on aerial thermal cameras,such as the DJI Zenmuse XT or the AgCon Aerial D1000, they are able to spot cows under forest canopies and distinguish animals from other heat sources. The cameras are also effective in detecting nearby predators.Using GPS, updates on each animal can be transferred to a Cloud platform. One of the best uses of drones is that these drones can fly a quick round of the cattle field at any time and anyone can easily review the video made by drones to check the number of the cattle. These drones help farmers to keep thieves away from the cattle field as they can be easily detected by thermal drones. For managing livestock, ranchers have adopted airframes to both herd cattle and monitor animals. Thanks to the ability to direct the aircraft's prop wash, drones are effective herding tools that let ranchers precisely herd the cattle over their property. The DJI Mavic Enterprise has a feature that lets the machine record sounds such as a sheepdog's bark then play them over a loud speaker projected across a paddock.

In 2015, Irish farmers Declan and Paul Brennan began to use a *drone* they called called *Shep* to herd their 150 sheep on their farm on the Carlow Laois border located 84 km (52 mi) south-west of Dublin.

Fertilisation

An example of use in fertilisation is the French drone manufacturer AIRINOV, whose precision fertilising had a positive impact on farmers' income and the environment. A three-year study on wheat crops showed that thanks to fertilisation advice provided by information from a drone, French farmers increased their gains by • 69 per hectare (2.5 acres).

Acquah Meyer Aviation is the brainchild of Ghanaian entrepreneur Eric Acquah and his wife Tracey Meyer, the company's Chief Operating Officer. While living in Germany, the duo used to see Ghanaian produce on the shelves at their local supermarkets. But they suddenly saw a drastic decrease, linked to the European Union's 2015 two-year ban on five Ghanaian vegetables. Farmers were not sufficiently spraying their crops, an agronomist told them, which had led to infestations. Determined to see Ghanaian exports back in European supermarkets, they launched their company in 2017. In Adeiso, Ghana, Daniel Asherow eyes a large white drone as it buzzes above his rows of pineapple plants, methodically spraying fertilizer into green stems that will soon produce juicy fruits for export. The drone hovers a few feet in the air, covering in 15 minutes the same ground that usually takes five workers an hour.

Irrigation Management

Previous chapter recounts the history of electrically-powered irrigation machines fixed on the ground. The agriculture industry is using drones to increase watering efficiency and to detect possible pooling, soggy spots or leaks in irrigation. Drones paired with thermal cameras are ideal in this area as they are able to detect and see from above what humans cannot from the ground. As the need for more efficient water usage increases, having a drone that is able to track and monitor irrigation is essential. With thermal and conventional cameras, drones are able to spot water pooling. With larger farms, having the ability to have a bird's eye view of what is being watered and at what time, allows farmers to more effectively use water resources.Drones can also irrigate. Equipped with hyperspectral, multispectral or thermal sensors, they can identify which parts of a field are dry or need improvements. Additionally, once the crop is growing, drones allow the calculation of the vegetation index, which describes the relative density and health of the crop, and show the heat signature, the amount of energy or heat the crop emits. On one crop alone, it is estimated that leak-detecting drones could save enough water to sustain more than 500 families for an entire year. Drone irrigation could reduce water consumption in farming by 90%

Pollination

About three-quarters of global crop species, from apples to almonds, rely on pollination by bees and other insects. But pesticides, land clearing and climate change have caused declines in many of these creatures, creating problems for farmers and for the rest of humanity. Pollination is needed for reproduction in flowering plants. Male flower parts, or stamens, produce pollen that fertilises female parts, known as pistils, to make seeds. In self-pollinating flowers, the stamen sheds pollen directly onto the pistil. Cross-pollination, however, requires the transfer of pollen from one plant to another. This mostly relies on pollen becoming stuck to the bodies of bees and other insects when they feed on flowers, and then being deposited on the next plant they visit. It has advantages over self-pollination, in that it increases genetic diversity and improves the quantity and quality of crops

One major challenge for the electrical agricultural airborne machine has been to save a species or to attempt to replicate its role in Nature. The honey bee (*Apis Mellifera*), which pollinates nearly one-third of the food we eat, has been dying at unprecedented rates because of a mysterious phenomenon known as Colony Collapse Disorder (CCD). The situation is so dire that in late June, 2014 The White House gave a new task force just 180 days to devise a coping strategy to protect bees and other pollinators. The crisis is generally attributed to a mixture of disease, parasites and pesticides. Inspired by the biology of a bee, researchers led by engineering Professor Robert Wood at the Microrobotics Lab of the Wyss Institute, Harvard University, began developing Autonomous Flying Microbotsaka *RoboBees*, man-made systems that could perform myriad roles in agriculture or disaster relief. A *RoboBee* measures about half the size of a paper clip, weighs less than one-tenth of a gram (3/100 oz!), and flies using "artificial muscles" compromised of materials that contract when a voltage is applied. To construct *RoboBees*, researchers at the Wyss Institute have developed innovative manufacturing methods, so-called Pop-Up micro-electromechanical (MEMs) technologies that have already greatly expanded the boundaries of current robotics design and engineering. A *Robobee* can lift off the ground and hover mid-air when tethered to a power supply. After two years of R&D, in 2016 the Wyss team announced that their *Robobees* can now perch on objects from any angle, using an electrode patch and a foam mount that absorbs shock to perch on surfaces and conserve energy in flight, like bats, birds or butterflies. The robot takes off and flies normally. When the electrode patch is supplied with a charge, it can stick to almost any surface, from glass to wood to a leaf. To detach, the power supply is simply switched off.The new perching components weigh 13.4 mg, bringing the total weight of the robot to about 100mg (3/100 oz), similar to the weight of a real bee.

But they still need to be able to fly on their own and communicate with each other to perform tasks like a real honeybee hive is capable of doing. The researchers believe that as soon as ten years from now these *RoboBees* could artificially pollinate a field of crops, a critical development if the commercial pollination industry cannot recover from severe yearly losses over the past decade. *RoboBees* will work best when employed as swarms of thousands of individuals, coordinating their actions without relying on a single leader. The hive must be resilient enough so that the group can complete its objectives even if many bees fail.[2] The new generation hybrid *RoboBee* can fly, dive into water, swim, propel itself back out of water and safely land. New floating devices allow this multipurpose air-water microrobot to stabilize on the water's surface before an internal combustion system ignites to propel it back into the air.

This latest-generation *RoboBee*, which is 1,000 times lighter than any previous aerial-to-aquatic robot, could be used for numerous applications, including agriculture.

Another approach has been taken by Anna Haldewang, a 24-year-old industrial design student at Savannah College of Art and Design (SCAD) in Georgia, USA. Haldewang created 50 designs of a bee drone before landing on the final model, *Plan Bee*, which does not resemble a bee at all. The drone consists of a foam core, a plastic-shell body and two propellers. There are also six sections of the drone that meet at the bottom, all of which have tiny holes that let the machine gather pollen while it hovers over plants. It can then release the pollen at a later time for cross-pollination. Haldewang noted that *Plan Bee* is still in its early stages, but she has filed a patent for the technology and design. Its application in backyards as a teaching tool has potential, but the drone could conceivably be used in large-scale farming, even in hydroponic farming.

A drone that can pollinate flowers may one day work side by side with bees to improve crop yields. Eijiro Miyako at Japan's National Institute of Advanced Industrial Science and Technology and his colleagues, have used the principle of cross-pollination in bees to make a drone that transports pollen between flowers. The manually controlled drone is 4 centimetres (1.6 in) wide and weighs 15 grams (0.5 oz). The bottom is covered in horsehair coated in a special sticky gel. When the drone flies onto a flower, pollen grains stick lightly to the gel then rub off on the next flower visited.

In experiments, the drone was able to cross-pollinate Japanese lilies (*Lilium japonicum*). Moreover, the soft, flexible animal hairs did not damage the stamens or pistils when the drone landed on the flowers. Miyako says the team is now working on developing autonomous drones that could help farmers to pollinate their crops. GPS, high-resolution cameras and artificial intelligence will be required for the drones to independently track their way between flowers and land on them correctly, though it will be some time before all that is in place.

Saul Cunningham at the Australian National University in Canberra says that using drones to pollinate flowers is an intriguing idea but may not be economically feasible. "If you think about the almond industry, for example, you have orchards that stretch for kilometres and each individual tree can support 50,000 flowers," he says, "So the scale on which you would have to operate your robotic pollinators is mind-boggling."

Australian technology start-up Bee Innovative with extensive experience in tracking honeybees in real time for precision pollination in Australia, and the University of North Dakota's global leadership in unmanned aircraft systems (UAS) are partnering to unlock an entirely new market for agricultural drones in the United States. The collaboration will focus on advancing the machine-vision-capability of Bee Innovative's current drone platform, *BeeDar* which is used by Australian farmersto track bee movements and pollination patterns in real-time and has delivered 20 percent increases in crop yields and returns for farmers season to season.

The grocery supply chain Walmart has filed six patents for drones that automate farming, including one that would identify pests, one that would pollinate plants, and another that would monitor crop health.

In April, 2019 mass application of drones for artificial pollination helped fruit farmers in Korla, in northwest China's Xinjiang Uygur Autonomous Region to take advantage of a two-week full-blossom period of fruit plants to grow premium fruits. In the renowned plantation for the Korla fragrant pear, the full-blossom period is less than two weeks. Korla has about 36,666 hectares (90,604 acres) of pear trees. Fruit farmers prefer artificial pollination instead of bee pollination, which can help prevent plant diseases from spreading.The flight route of the drone is set automatically after the operator confirms the boundary of gardens with an error range less than 5 cm (2 in). The drone takes off and returns by itself. XAG has1,950 drones in agricultural use in Xinjiang, which has occupied a lion's share of the market in the region.

Traditionally, date production is a particularly labour-intensive crop to grow. Workers must touch a tree probably 12 times over the course of the season. The date industry has been using sidewalk leaf blowers to pollinate the trees. In 2015, students from the University of Arizona started testing the idea of drone pollination using a nylon stocking. They hung the stocking under the drone and the rotors kicked up the pollen in the stocking and pollinated the date trees. After seeing success with this, a team of eight engineering students made this their senior project to build a mechanism a little fancier than a nylon stocking.

This was done with a 3D printer that would dispense pollen at the command of a control system, so it goes from one tree and dispenses the pollen and then onto the next tree so it is a savings of pollen and of labour. During 2019, backed by the Yuma Center of Excellence for Desert Agriculture, the Bard Date Company teamed up with the University of Arizona for on-site trials.

The above employment of agricultural drones is only a selection of a burgeoning crop of applications, so reducing the workforce to trained operators.

Endnotes

[1]Haye Kesteloo, "Over 400 DJI drones in world's largest agricultural drone fleet" dronedj.com, February 25th 2019.

[2]Dina Spector, "Tiny Flying Robots Are Being Built to Pollinate Crops Instead of Real Bees," Business Insider July 7, 2014; Alexia Erickson, "Robotic Bees are now being built to pollinate crops instead of real bees," NextStory, October 5, 2016.

6

Driverless Agricultural Future: The Robotic Farm

In 1920, a 30-year-old Czech writer Karel Èapek, wrote a play called "R.U.R." (*Rossumovi Univerzální Roboti* or Rossum's Universal Robots). For this dystopian work about a bad day at a factory populated with sentient androids, Èapek called the androids "roboti", as suggested byhis brother Josef, a painter and writer. It was premiered in Prague on 25th January, 1921.

Èapek was admired for his opinions:

> My dear Miss Glory, Robots are not people. They are mechanically more perfect than we are, they have an astounding intellectual capacity, but they have no soul.

But surprisingly he was keen on gardening!

> ... and not too much; that there may be plenty of dew and little wind, enough worms, no plant-lice and snails, no mildew, and that once a week thin liquid manure and guano may fall from heaven. Amen. ... *The Gardener's Year*, 1929

Although the Èapeks lived to see the beginning of robotic prototypes, they could not have envisioned that a century later, words like Mowbot, AgriBot, DroneBot and FarmBot would come into agricultural vocabulary.

In 1935, Otfrid von Hanstein's science-fiction short story, "The Hidden Colony" was published in *Wonder Stories* comic:

> In front of me, everything seemed alive; the huge plough, which had been standing near by the house without my having observed it in the night, seemed to shake itself, turn around, swung its arms, creaked - and began to move back southward whence it and we came, tearing up the vegetation, ploughing the ground beneath the wrenched-out plants and shrubs.
>
> All around me were machines, busily at work, machines that threshed and winnowed grain... A picture of machines and no man to control or watch them!
>
> Machines that seemingly with full consciousness walked out into the fields to do their daily work. And even now there was no living being among them save myself... Had these machines in some incredible fashion been provided with brains?

In the sci-fi film "Star Wars, Episode I, The Phantom Menace" written and directed by George Lucas in 1999, moisture farmers on dry scorched planets like Tattoine use GX Water Evaporators manufactured by Pretormin Environmental.

Farming is hard work. About 35 percent of China's labour force is in agriculture (compared to 2.5 percent in the U.S.). There are 425 million agricultural workers (200 million farming households) in China. According to Registrar General of India & Census report 2011, the total number of farmers or cultivators in India is 118.7 million with another 144.3 million agricultural workers/labourers which makes up almost a third of the total rural population. The more than 2 million farms in the U.S. employ roughly 925,000 people to perform tasks such as planting, seeding and inspection, contributing to total production expenses of $350 billion in 2017. Farming worker shortages are getting worse. In a survey by the California Farm Bureau Federation in 2018, 55 percent of the 762 farmers surveyed said they had experienced employee shortages. Fewer people are working on farms. But without an increase in agricultural productivity of at least 60 percent, the world might not grow enough for the projected global population by 2050. That is why researchers are now beginning to tackle this problem with robots.

The idea of a driverless farm tractor has been around since as early as 1940, when Frank Winston Andrew of Palmyra, Illinois invented his own. To guide his driverless tractor, a barrel or fixed wheel would be put in the centre of the field and around it would wind a cable attached to a steering arm on the front of the tractor. In the 1950s, Ford developed a driverless tractor that they called *The Sniffer* but it was never produced because it could not be operated without running wire underground through the field. In 1981, Timothy R. Pryor of LMI Technologies Inc. took out a patent for a "Robot Tractor" system, an electro-optical and microcomputer-based method and apparatus for automatically guiding tractors and other farm machinery for the purpose of automatic crop planting, tending and harvesting. Also disclosed were means for automatic picking, excavating and other off-road uses in relatively confined areas.

There were no major advances in driverless tractor technologies until 1994 when engineers at the Silsoe Research Institute in Bedford, England developed the picture analysis system, which was used to guide a small driverless tractor designed for vegetable and root crops. This new tractor could even handle slight headland turns. In 2005, Nicholas D. Tillet and Tony Hague founded the company Silsoe Ltd., to provide automation technology expertise for the agricultural industry that would otherwise have been lost on the closure of Silsoe Research Institute. The company's main focus continued to be highly innovative research and development in computer vision based guidance and control. These research findings were made available from a licensee Garford Farm Machinery as Robocrop Electric.

During the past twenty years, a team of researchers at Carnegie Mellon University in Pittsburgh, Pennsylvania have been researching into sensors, artificial intelligence (AI) and robots for agriculture to create a fleet of mobile field robots under the headline FarmView. In 2000, the project leader George A. Kantor started working with a project that was using wireless sensor networks to collect environmental information in container nurseries. The network measured soil moisture, temperature, sunlight and several other parameters, and Kantor's team developed methods to use that information to help make better irrigation scheduling decisions. That work eventually led to a large project sponsored by the U.S. Department of Agriculture to develop intelligent irrigation. Borrowing ideas and technologies from CMU's successful DARPA Urban Challenge self-driving car, Carnegie Mellon developed an autonomous utility vehicle that could drive up and down the rows of an orchard to perform various tasks, such as mowing and harvest assist. They then began collaboration with Clemson University on a project to use robotic phenotyping to accelerate the breeding of sorghum as a biofuel crop. They provide breeding and genetics expertise and conduct field work at research farms in South Carolina and Mexico, where FarmView deploys its robots to collect the necessary data and work together to determine which measurements to take and to interpret the results. The sorghum-stabbing robot project puts technology to use in the field of crop breeding. Traditionally, crop breeders combine different strains of a crop (there are 40,000 varieties of sorghum) to create many different "children" crops of new strains, which they have to grow to test out. This allows them to breed new versions of the crop that are more resistant to disease, or to drought, or have a higher yield. This is a timely process and tracking all of the different new strains, how they are developing and what their attributes are, is difficult. Using FarmView speeds the process up significantly. FarmView is also working with Cornell and the University of California Davis (UCD) on a project to increase efficiency and quality in grape production by actively managing vine balance. The group also worked with several land grant universities led by the University of Maryland and including the University of Georgia, Colorado State University and Cornell to develop an intelligent irrigation control system for nursery and greenhouse crops.

The main sources of electrical power for robots are batteries. The type of battery that is used for a robot varies depending on the safety, life cycle, and weight. Lead acid batteries are common, as are silver cadmium batteries. Rechargeable batteries and primary batteries are both used; batteries that are not rechargeable are generally more powerful. Other options for robotic power sources are thermoelectric generators, which convert heat directly into electricity; fuel cells, which are similar to batteries except fuel and oxidants are continuously supplied; supercapacitors, which store high energy as a charge built up on plates; and tethers, which connect the robot to the power supply. The tether option entirely removes the power source from the robot which saves weight and space, but since the robot constantly has to be connected, it can be a nuisance as well.

From 2011, Kraig Schulz and Terry Anderson of the Autonomous Tractor Corporation of Saint Michael, Minnesota began to develop technology to save farmers 50% on their equipment costs by retrofitting their existing tractors with electric drive technology. Schulz grew up in Minnesota and earned an MS from the School of Agriculture at the University of Arizona. A former partner at Ernst and Young, he spent four years directly working with agricultural communities in Mali, West Africa. Anderson grew up on a Minnesota grain farm and spent several decades as a serial tech entrepreneur before he sold Ancor Communications, a manufacturer of fibre channel network products for computers, in 2000 for $2 billion. Their prototype, put together by a team including Bob Cornelius, Frank Artner, although a blocky champagne-coloured engine atop treaded rubber tracks, drew crowds when it debuted at the 2012 Big Iron Farm Show in Fargo.

Withe Drive, one of ATC's products and the only diesel-electric drivetrain for tractors, ATC aims to solve many of the cost of ownership issues. In parallel, they designed AutoDrive, a laser-radio navigation system (LRNS) that addresses the limitations in today's GPS-based systems for driver assistance. Using the Auto Drive management system, vehicles "learn" their way around fields after a preliminary run with the farmer in the cab. Farmers can then freely switch from manual to autonomous steering and let the tractor complete the field work independently. ATC's guidance system relies on proprietary laser and radio systems to navigate fields and remain within boundaries, while broad-spectrum sonar pulses help the vehicle avoid obstacles from 9 metres (30 feet) away. If a tractor loses its sense of position, identifies a problem in the engine, implements or motors, or if sensors notice a power line or stray barnyard critter in its path, the failsafe stopping mechanism kicks in, ceasing operations instantly. The two patented products, eDrive and AutoDrive,are initially sold in the after-market as retrofits.

In 2013, a team led by Simon Blackmore at Harper Adams University near Newport, in Shropshire England unveiled their AET *Pomona*, which they had developed as part of a project to create sensors to be placed on an autonomous orchard tractor, to gather data, analyse and present information to the farmer prior to irrigation and harvesting during the growth season. It was entitled

"Usability of Environmentally Sound and Reliable Techniques in Precision Agriculture" (USER-PA)and brought together scientists from Israel, Germany, Turkey, Switzerland, Greece, United Kingdom, Italy and Denmark. The project combined expertise from universities, research institutions and industry in the fields of plant physiology, sensor technology, robotics and automation as well as decision support systems. *Pomona* was created from modifying an existing tractor to enable it to be controlled purely by a computer. The Harper Adams team removed the seat and steering wheel as they were not needed anymore.

Inspired by the work of the team at Harper Adams University, Ben Scott-Robinson founded the Small Robot Company in Portsmouth, England. Scott-Robinson had worked on the digital transformation of the Ordnance Survey maps, and has been involved in shaping the user centred design technology behind the autonomous vehicles soon to become commonplace on roads. Scott-Robinson was joined by Joe Allnutt, who had worked with him on digitising OS, running the tech labs to develop hardware as well as AI systems to manage the data. They welcomed Sam Watson-Jones of Newport, Shropshire, a fourth-generation farmer with field experience. They have also partnered with Cosmonio, a UK-based start-up that specialises in developing AI systems that automate the process of extracting visual information from images using self-learning algorithms.

Initially, the Small Robot Company team developed *Rachel*, an agile four-wheel monitoring robot with a bright orange 3D-printed body and beefy all-terrain wheels, whose task is to obtain a plant-by-plant view of the fields, which enables digitisation. *Rachel* is quite small 50 x 60 x 40 cm (20 x 24 x 16 in), weighs around five kilos (11 pounds) and is fitted with cameras. It runs backwards and forwards, over plots of land and collects data about the plants present. It monitors

Ben Scott Robinson with the Small Robot Company's *Tom* prototype (Small Robot Company).

Small Robot Company's *Harry* being unveiled at the REAP 2018. Sam Watson Jones and Belinda Clarke (StillVison Photography).

every single plant in a field and identifies the difference between crops and weeds like black-grass or brome. *Rachel* was renamed *Tom*.

*Tom (*ex-*Rachel)* is part of their "Farming as a Service" (FaaS), a new way to grow crops. The idea of FaaS is that the farmer does not actually buy a robot, but contracts in the service, paying a rate per area. FaaS uses *Wilma*, a data storage and management system that sits in the Cloud draws all the information from *Tom* the scout, combines it with weather data, crop and field history, plugs into pathogen-monitoring and similar services, and delivers direction to *Dick* to carry out remedial work based on this plant-level data. *Dick* then accurately targets weeds for micro-spraying and mechanical weeding, then also delivers fertiliser directly to soil around roots, instead of wasteful blanket spraying, and uses laser or micro-spray chemicals to kill weeds. *Harry* will plant seeds in the ground, at a uniform spacing and depth, allowing them to grow effectively, eliminating the need for tractors to plough furrows. Their names are taken from the film "Blade-Runner", a 1982 science fiction film directed by Ridley Scott. The Small Robot Company has filed several patents for these farmbots and it is running some small tests with farmers. However, they will not be commercially ready for use until autumn, 2021. The disruptive plan of the Small Robot Company is to own and administer a nationwide fleet of Toms, Dicks and Harrys, ensuring they are all running effectively and carrying out any maintenance, upgrades or replacements required.

In 2016, Case IH presented their Autonomous Concept Vehicle. They then field-trialled a small fleet of autonomous Steiger Quadtrac tractors pulling a True-Tandem disk harrow or Ecolo-Tiger disk ripper, in collaboration with one of the largest carrot producers in North America, Bolthouse Farms. The same year New Holland, who also debuted their NHDrive in 2016, partnered with E. & J. Gallo Winery, the largest family-owned winery in the world, to begin trialling their technology on T4.110F vineyard tractors.

In India, Escorts Agri Machinery has unveiled an electric tractor concept under the Farmtrac label.It was announced as one of the country's first AETs, developed in collaboration with seven technology giants namely Microsoft, Reliance Jio, Trimble, Samvardhana Motherson Group, WABCO, Bosch and AVL. According to the company, the partnership between these technology corporations would help the company to develop a range of farm machines with electric transmission, autonomous applications, remote vehicle management, data-based soil and crop management, and sensor-based guided farm applications.According to the Escorts, Indian agriculture and farming practices require extensive mechanisation and precision based agro solutions for maximum output. For this reason, Escorts has collaborated with AVL for electric transmission technology; with Trimble for sensors, control, water level management system and autonomous e-steering; with Samwardhana Matherson Group for smart interface Cabins & Care Plus; with WABCO for vehicle controls and automation technology and with Microsoft for its Cloud and AI technology enabling precision agriculture capabilities. Reliance Jio will help to enhance farm machinery life cycle with a network platform by providing service and spare parts across the country. The Bosch start-up, Deepfield Robotics would handle emission readiness. Deepfield robots navigate plant rows, sense the plants and send the data to the farmers to help optimise seed breeding. If equipped with a "weed puncher", the robot can literally drive weeds into the ground. Deepfield Robotics also provides smart sensors that can be positioned in the fields. Resulting networks are already deployed in farms to monitor soil conditions for asparagus

Mahindra and Mahindra Ltd. also announced the *Hulk*, their version of the AET, working with Kaustubh Dhonde at AutoNXT Automation, a completely bootstrapped Mumbai-based start-up. The *Hulk*, a 30hp tractor that can run for 150 km (93 mi) on each charge, would be shared by communities of farmers who own smaller pieces of land.

Another builder of small land-based robots is Jaisimha Rao, who founded TartanSense in Bengaluru, Karnataka, India in 2015. Rao was previously with BlackRock, managing portfolios of securitised products. He is an engineering graduate from Carnegie Mellon University. In an effort to bring more data to Indian farmers and improve their decision making, Rao started by deploying drones as an excellent way to gather data at scale, and it was initially transformational to look at farms from the sky. But then he ran into a wall: the imagery did not go far enough. Rao's customers wanted more than just information about their fields, they wanted him to help them solve the issues identified by the drones. Rao pivoted and founded TartanSense, to build small land-based robots for small farms (and smallholder farmers), initially focused on pesticide application. The TartanSense agribot is a semi-autonomous rover, which travels across the fields with a downward facing camera. The AI/ML system identifies pests using computer vision and then targets precision sprayers on them. The rover and its accompanying software will also gather land-based data at a much higher resolution than from drones, generating reports with actionable insights for the farmer. India is the largest cotton grower in the world, with 12 million hectares (30 million acres) of cotton farms, which has made it a good crop for TartanSense to start with. The start-up is also looking at soybeans, shortlisting these crops due to their high value and the high cost of their inputs.

With the slogan "Think outside the cab!" in 2017, Igino Cafiero and Aubrey Donnellan founded Bear Flag Robotics of Sunnyvale, California to develop an AET which while operating in an orchard with a mass of canopy cover, cannot use GPS. They are well qualified for the task. Cafiero grew up in a tech family, obtaining his Bachelor's Degree in electrical and computer engineering at Carnegie Mellon University, then his Masters' Degree in electrical engineering at Stanford University. He has built several race cars including a Baja race truck and has worked on all types of engines. His experience with both software and physical machines fuelled his enthusiasm for industrial automation and led him to build the first Bear Flag prototype in his garage. While working for Phantom Works, the R&D division of Boeing Satellite Development Center, Donnellan designed and implemented advanced robotics and manufacturing solutions to support the company's first satellite manufacturing line. Pulling research from adjacent industries such as video gaming, his team prototyped one the first augmented reality solutions to aid in complex assemblies. In 2018, Bear Flag Robotics announced its plans to build.

In 2017, Kubota of Osaka, Japan also developed an autonomous tractor. Called the Agri Robo, the diesel tractor is based on a SL60A and brimming with sensors and cameras. Kubota has also provided a solution for farmers who want more electrical power than that delivered by the traditional 12-volt socket. The e-Power platform, installed at the rear of Kubota tractors, is designed to deliver up to 10.5 kW. The outlet is placed just below the PTO, which allows a mixed feed, for example for a fertilizer distributor. The platform is Isobus compatible and integrates the TIM (Tractor Implement Management) function, that is to say the control of the tractor by the tool. Thus, the equipment can control the shutdown of the power supply. In 2018, within the framework of Kubota's ongoing growth in Europe, Kubota Corporation has announced that it would invest •55 million in a new R&D centre in Europe. The new site would be located in Crépy-en-Valois, France and will be fully operational in 2020.

Between 2008 and 2011, Bakur Kvezereli was Secretary of Agriculture at Tbilisi, Georgia, responsible for agriculture policy, food safety and agriculture infrastructure including agricultural machinery. *Bakur's single mission was to allow farmers to sustainably operate with environmentally-friendly and efficient 100% electric farm vehicles. Having emigrated to the USA, in 2017 Kvezereli co-founded a* specialty food marketplace Maiaki and another start-up Ztractor in Palo Alto, California to design and manufacture the world's first autonomous electric tractor (AET) to help fruit and vegetable farmers. He calls his models *Bearcub 24*, *Mars 45* and *Superpilot 125*. In 2019, a *Bearcub 24* was trialled on a strawberry farm in Norway.

In Japan, Yanmar Co. Ltd. of Osaka, in collaboration with Hokkaido University, developed two lines of autonomous tractors Robot Tractor and Auto Tractor. Robot Tractor can work alone or in a group, using the Smartpilot ICT (Information and Communications Technology) system for autonomous movement. The Auto

Tractor model can perform work with high accuracy and minimal human intervention. This machine is equipped with a GPS control system, and if necessary it can be converted into a Robot Tractor. In addition to advancing forward, "Auto Mode" allows the tractor to automatically drive in reverse, stop and execute turns. In "Linear Mode" while certain driving manoeuvres such as turning are performed by hand, the tractor can travel back and forth on its own when cultivating land. Switching between these two modes is possible in the 2-Series, allowing you to customize it for different operators, or according to skill of the individual and the work itself. At present these are diesel-fuelled, but this may change.

In China, Henan Intelligent Agricultural Machinery Innovation Centre at Luoyang unveiled *Epoch* AET which began trials in January 2019.

Also on trials is John Deere's AET, developed by a team including Steve P. Robisky, Manager Strategic Planning of "Farm of the Future" and Troy K. Maddox, Senior Designer UX/ID. The modified John Deere *6210R* tractor uses a robot arm to pay out and retract its GridCON, a 1 km (0.6 mi) long onboard extension cord from the machine to the field perimeter, which provides up to 400hp (300kW) of power in total. Pairing autonomous performance with the extension cord has the potential to lower the capital cost of purchasing the vehicle by removing the batteries. At the same time, the autonomous capability of the vehicle ensures that it will not run over its own power cord and reduces wear and tear on the cord itself. Compared to battery-powered equivalents, the prototype delivers around 50% lower machine and operating costs.

John Deere's Autonomous Electric Tractor uses a robot arm to pay out and retract its GridCON, a 1 km long onboard extension cord from the machine to the field perimeter, which provides up to 400hp (300kW) of power in total. Is this the return of the cable plough, invented in 1879? (John Deere).

In October 2019, at a dealer meeting in Valencia, Spain, John Deere unveiled a driverless electric tractor with a rated power output of 500 kW (approximately 680 horsepower). The concept combines what resembles the front half of a rubber tracked tractor connected via an articulation point to a set of discs. The tractor's rear axle and cab have gone, and the implement appears to be carried on the rear wheels for transport. The discs can be swapped for other kit. Perhaps most revolutionary of all is AGCO and Fendt's FutureFarm Project *Xaver*, using Mobile Agricultural Robot Swarms (MARS). The use of a large number of small, identical robots operating in a swarm enables smooth running of the job, even in the event of the failure of a single unit. The "mother ship" arrives and the small agbots disembark to carry out their tasks.

Fendt's FutureFarm Project *Xaver*, deploys Mobile Agricultural Robot Swarms (MARS) enabling smooth running of the job, even in the event of the failure of a single unit. The Mother Ship arrives and the small agbots disembark to carry out their tasks. (Fendt).

Fendt and Braun Maschinenbau teamed up to equip the 200 Vario into an Automated Vehicle using implemented guidance system using a satellite-based system via GNSS or through ultrasonic sensors for use in, for example, wine growing.

A similar approach is with Rabbit AETs. For thirty years, David Miklosi worked with emerging technologies, starting with the launch of seven of the first modern electric vehicle fleets to customers then with the transition of a Stirling-cycle (heat) engine from R&D to commercialization. Since 2016, with entrepreneur Zac James, Miklosi has developed Rabbit Tractors in Ann Arbor, Michigan, USA: these are small autonomous swarm-enabled machines and multi-purposed farm equipment with a focus on core IP areas unique to agricultural environments. Because current tractor size makes them inefficient to transport, they are geographically constrained by where they can be efficiently driven at low speeds. Multiple Rabbits can be loaded onto standard trailers and transported further distances, at greater speeds on highways. Swarm connectivity makes Rabbits modular, allowing farmers to buy the exact amount of equipment for their operation and then add more equipment as their operation grows. They measure 1.5 m long by 0.9m wide (5ft x 3ft), weigh 114 kg (250 lbs0, with a 119hp 4wd hydrostatic drive system and 20hp PTO output. Unlike gear transmissions, hydrostatics has a continuous power curve without peaks and valleys, and they can increase available torque without shifting gears. With their unique size, design and patent-pending features, Rabbit Tractors are the only AET capable of pulling implements, planting, as well as high-boom crop spraying.

Having grown up on a farm in Iowa and studied agriculture and business at Iowa State University, in 2011 Colin J. Hurd founded Agriculture Concepts. With the help of the College of Agriculture and Life Sciences Agricultural Entrepreneurship Initiative, Hurd launched and patented TrackTill, designed to eliminate the compaction caused by large row planters. TrackTill consists of vertical rolling tines, which slice the soil to relieve compaction up to 30 cm (12 in) deep. The tines pass through the soil subsurface without disturbing the seedbed. He then promoted Caden Edge, a patented welding process that enables plough sweeps and points to last 3 to 5 times longer than standard equipment. This was followed across Midwest by Aquacheck soil moisture monitoring technology to accurately show rainfall events and soil saturation levels remotely. Acquachek capacitance probes measure soil moisture and temperature simultaneously every 10cm (4 in) depth. A capacitance sensor sends an electric current into a surrounding volume of soil to measure its moisture content (a measure of the dielectric permitivity of the soil), these are then relayed wirelessly to an internet Cloud. In 2015, Hurd founded Smart Ag in Ames, Iowa inventing and patenting AutoCart, a software-controlled autonomous tractor for grain harvest. Total installation by a dealer can be completed in less than eight hours and once installed, the farmer can run the AutoCart app on their tablet. They type in a destination on a field map, drop a pin and the tractor heads for the destination. When the combine is full, the farmer can request synchronization and the tractor drives alongside, allowing for unloading. The Smart Ag kit for AutoCart comes with an automation kit, hardware to connect machines to the Cloud and tractor, the software and the overall farm, field and machinery platform for autonomous farming.

Troy Fiechter of Rogo in West Lafayette Indiana, has developed Smart Core an autonomous soil sampling vehicle that uses robotics to gather samples with a consistent depth, and with proximity repeatability to ensure accurate sampling year after year. It operates at a 202-hecatre (500-acre) capacity without stopping and can hold up to 200 samples. The SmartCore uses a Bobcat tracked skid steer chassis as the base, and a hydraulic-powered auger bit is used to collect samples. The bit is configured so that no soil can escape the core, ensuring each sample is pure and accurate. The extraction arm is rotationally received on a housing and movable to an inverted position for depositing the quantity of extracted soil into a packaging assembly for collection, labelling and storage of the individual samples. While the machine is autonomous, an operator helps open up a field and evaluates potential trouble sites, such as ditches or other obstacles. While the Smart Core gathers soil, a Rogo employee can package soil samples for delivery to a soil lab of the farmers' choice. Covering 324 hectares (800 acres) per day, Smart Core autonomously sampled 8,094 hectares (20,000 acres) during the autumn of 2018.

Then there is Einride's *T-Log*. Robert Falck together with Linnea Kornehedand Filip Lilja of Stockholm, Sweden founded Einride to develop the self-driving *T-pod*, electric truck that is not just able to operate autonomously, but does not even feature a driver's seat. Without the cabin, more space is left for the drivetrain and freight. Instead, it carries a more compact suite of lidar, radar and camera sensors, feeding information about the environment to Nvidia's AI supercomputer. Powered by a 200 kWh battery, *T-pod* can cover approximately 200 km (125 mi) on a single charge. Over that distance, it has space for 15 standard pallets, equal to 15 m^2(160ft^2). Its maximum weight will be around 20 tonnes (20 ton). In October, 2017 Einride released a statement explaining how it was going to support Lidl in its transition towards electric and autonomous transport. The supermarket chain has a goal of reducing its emissions by 40% in Sweden by 2035. The company started a pilot programme using the *T-pod* by the third quarter of 2018. The vehicle has been driving around the city of Halmstad on the Swedish west coast, carrying real Lidl products and driving through Swedish traffic. At first the *T-pod* was monitored at a distance by a human driver, ready to take over the wheel when required. Total autonomy, at least at the start of the pilot, had not yet been reached, but Einride plans to slowly decrease human involvement, going from 1 to 10 trucks per remote driver. By 2020, two hundred *T-pods* will be transporting 2 million pallets on the route from Helsingborg to Gothenburg.

In July, 2018 Einride unveiled the *T-log*, with a stronger suspension for the logging industry. Acknowledging that computers cannot handle every on-road driving scenario yet, let alone unmarked lanes under heavy tree canopies, Einride will be able to drive its *T-logs* from a remote location using teleoperation tech provided by Phantom Auto. That is enough ambition to make an 18-wheeler ride low, but Einride plans to put vehicles on the road later this year, starting with a *T-pod* truck carrying tyres between two depots belonging to DB Schenker, one of the world's largest logistics companies, in Jönköping, Sweden.

One of the key players in the development of agricultural robotics is Salah Sukkarieh, born in 1973 in Sydney, Australia of a family with Arabic origins.Having completed his degrees at the University of Sydney, the title of his PhD being "Navigation Systems for Autonomous Robots", in 2004 Sukkarieh returned to join the staff. From 2007, he became Director of Director Research and Innovation at the University-based Australian Centre for Field Robotics

Einride of Stockholm, Sweden have produced the self-driving T-Log, which cannot just operate autonomously, but does not even feature a driver's seat. Einride will be able to drive its *T-logs* from a remote forest location, using teleoperation tech provided by Phantom Auto (Einride).

(ACFR), leading the strategic research and industry engagement programme in what has become the world's largest field robotics institute with 120 researchers and engineers. During this period of time Sukkarieh worked on many projects: he began with automated weed detection and moved on toflight planning systems for commercial aviation and automation systems for mining. But his largest effort was on agriculture robotics.

By 2014, Sukkarieh et al. had produced a giant robotic weed-remover research platform called the *Ladybird* on account of the curved solar panel on its back. For this, Salah was awarded the NSW Science and Engineering Award for Excellence in Engineering and Information and Communications Technologies. RIPPA followed (Robot for Intelligent Perception and Precision Application), an operational prototype that would work on commercial farms. In 2016, ACFR developed *Shrimp* to herd groups of 20 to 150 dairy cows. Modifications enabled keeping an eye on their health, and checking that they had enough pasture to graze on. Sick and injured animals would be identified using thermal and vision sensors that detect changes in body temperature and walking gait. Trials took place on several farms in central New South Wales. As is reported in Chapter 6, drones are also carrying out similar duties – identifying sick animals and herding them into pens.

In 2016, Salah Sukkarieh (right) and a team at the Australian Centre for Field Robotics (ACFR) developed the low-cost, remote-controlled *SwagBot* to focus on large-scale crop and livestock operations including weed removal and yield counting (Courtesy: Salah Sukkarieh).

By 2016, ACFR had produced the low-cost remote-controlled *SwagBot* to focus on large-scale crop and livestock operations including weed removal and yield counting. In Australian history, a swagman was a transient labourer who travelled by foot from farm to farm carrying his belongings in a swag (bedroll). An initial field trial in Bandung, Indonesia allowed Sukkarieh and his team to gain insight into the challenges faced by farmers in developing countries. Thanks to a philanthropic gift to the University of Sydney to look at developing low cost digital technologies that would support Australian farmers, ACFR progressed to the digital FarmHand to improve productivity on small farms in Australia. Digital FarmHand is designed to help farmers automate tasks and via simple smartphone technology to assess plant health, and through smart tools to conduct spraying and weed control, perform crop analytics and automate simple farming tasks. Like a tractor, Digital Farmhand uses a 3-point hitch mechanism which allows the attachment of various implements such as a seeder, sprayer and tine weeder. As many of the farms are located in remote areas, the robot is easy to maintain and uses basic manufacturing techniques enabling it to be repaired locally with off-the-shelf parts. The new technology was debuted in partnership with Greater Sydney Local Land Services. August, 2018 Sukkarieh and his team carried out trials of Digital FarmHand in Fiji and Samoa to find an inexpensive way for small-scale farmers in the Pacific Islands to increase crop yields, particularly of nutritious foods. In 2019, the University of Sydney agricultural robotics spin-out, Agerris, led by Sukkarieh, raised AUS $6.5 million to commercialise FarmHand and to prepare a cattle-herding AI version. Professor Sukkarieh, with over 500 academic and industry publications in robotics and intelligent systems, was nominated as NSW Australian of the Year for 2019.

Visionary Italian designer Lorenzo Mariotti of Rome has come up with a tractor concept for the future, dreaming up what he calls the H202 concept, a hydrogen fuel cell EAT (Electric Autonomous Tractor). Focused on time optimisation and cooperation with the user, the H202 includes technologies such as laser and OLED lights, advanced active safety and remote control. The H202's engine, equipped with fuel cell that produces electricity from hydrogen charged directly at the farm, using a plug or wireless inductive charging, whilst hydrogen can be obtained from methane produced with anaerobic digesters directly on the farm, reducing the overall footprint of the entire farm.

Harvestbots

Alongside Electric Autonomous Tractors travelling up and down fields and plantations, there are those robotic machines which harvest with great precision.

From 2001, James D. English and Chu-Yin Chang at Energid Technologies in Cambridge, Massachusetts began work on a premier commercial software development kit (SDK) and tasking framework that supports real-time, adaptive motion control. One of their products is a robotic citrus-harvesting system which can pick a fruit every two to three seconds. To achieve this, it uses multiple low-cost picking mechanisms organized into a grid. The gridded picking mechanisms are simple, with two degrees of freedom each for aiming and one degree of freedom in extension. The extending parts of the picking mechanisms have no actuators and no sensors, making them robust, easy to manufacture, and easy to replace. Organizing the picking mechanisms into a grid allows the six cameras used to locate fruit for picking to be organized into an integrated grid rigidly attached to the supporting structure. The Energid harvester was tested in a Valencia orange grove in Florida using a portable generator and air compressor. For field testing, the picking system was mounted on a four-axis hydraulic arm mounted on a diesel truck. The hydraulic arm allowed an operator to move the picker through a scanning motion over the surface of citrus trees, while the picking system automatically found and removed the oranges in the area that could be seen by the cameras.

Since 2009,Juan Bravo has been developing a strawberry-picking robot called *Agrobot*, at the Centro de Innovación Tecnología de la Agroindustria Onubense (Centre for Technology Innovation of the Agricultural Industry of Onuba), in Huelva, a port city in south-western Spain. SW6010, the prototype was ready and following further R&D, the latest*Agrobot* has up to 24 robotic arms working wirelessly as a team. Each one is fully independent from the other and each one has a camera. The flexible platform fits into any farming configuration. *Agrobot* could harvest around 8 hecatres (20 acres) in three days. *Agrobot* is now based in Los Angeles, California.

Harvest Automation Inc. in North Billerica, Massachusetts was founded in 2009 by former iRobot employees on the premise that there are many examples of work in the world that are repetitive, strenuous and dangerous and that can benefit from robotic automation. "Roomba" inventors Joseph Jones and Paul Sandin, and "Roomba" engineer Clara Vu, founded the company along with CEO Charles Grinnell. Harvest Automation's products are branded under the OmniVeyor product line. Harvest Automation's first product was the OmniVeyor *HV-100*, providing automated material handling for the nursery, hoop house and greenhouse market. When plants are young they are placed in close proximity to one another on fields in order to conserve resources such as land, water and pesticides. As the plants grow their containers must be spaced further apart to allow the plants to develop a uniform and consistent canopy. The task of spacing plants has historically been completed with manual labour.Working in small teams, OmniVeyor *HV-100* robots automate the task of spacing containerized plants, common to wholesale nursery and greenhouse operations.

In 2010, European Union Project FP7 named CROPS was set up to research nanotechnologies, materials and new production technologies for automation and robotic applications to sustainable crop and forestry management. The project was coordinated by Wageningen University & Research, Netherlands. Out of this, as part of the European Union's Horizon 2020 innovation programme, a sweet pepper-harvesting robotproject called SWEEPER was launched in 2014. SWEEPER involved 6 partners from 4 different countries. The consortium included fundamental and applied research organizations: Wageningen University & Research and Stichting Wageningen Research from the Netherlands; Umea University, Sweden; Ben-Gurion University of the Negev, Israel; Proefstationvoor de Groenteteelt; PSKW, Belgium: B&A Automation, Bogaerts, Belgium, an industrial company as system integrator; and a modern grower of sweet pepper from the Netherlands. The project concluded in October, 2018 with two demonstrations in greenhouses: one a commercial greenhouse and the other at the Proefstation voor de Groenteteelt in Sint-Katelijne-Waver in Belgium. To pick a single pepper took about 24 seconds, though the researchers said they purposefully slowed down the robot's movements for safety reasons. SWEEPER is also equipped with LED lights so that it can work regardless of the time of day, for about 20 hours/day. Still, the robot was far from perfect, with only 61 percent accuracy in picking ripe fruit. The SWEEPER robot was on display and presented during the IROS (IEEE/RSJ International Conference on Intelligent Robots and Systems) in October, 2018 in Madrid.

Harvest CROO (Computerized Robotic Optimized Obtainer Robotics) was founded in 2013 by strawberry grower, Gary Wishnatzki of Plant City, Florida and robotics engineer, Robert Pitzer, formerly a submarine qualified U.S. Navy nuclear operator, Intel automation engineer, entertainment and STEM Large Scale Robot Competition producer. Their goal was to make a robot that could pick strawberries. Each successive prototype was called *Berry*. *Berry 5* was trialled out in Florida fields prior to commercialisation before the end of 2019.

In 2014, Tom Coen and a team in Heverlee-Leuven, Belgium began to develop *Rubion*, an autonomous robot that could delicately pluck strawberries straight from the bush and place them softly in a basket. They started up a company called Octinion. Using machine vision, *Rubion* can identify ripe strawberries and using a 3D printed 'hand' harvest the perfect fruit. Its vision can also estimate when the unripe fruit will ripen. Octinion has also developed a robot for treating crops with UV-C light to fight mildew and therefore reduce the number of chemical residues.

Due to the way that broccoli crowns grow, the green vegetables are difficult to harvest and therefore produce little income. To combat this, OSU created broccoli with crowns that grew above their leaves, but this new creation was left un-marketed because there were no harvesters specifically designed for this new kind of hybrid broccoli. From 2015, a team of researchers led by Carter Precourt at Crescent Valley High School in Corvallis, Oregon, USA developed *Broccoli Bot*, that would mechanically harvest this new kind of broccoli. Another Broccoli Bot was developed by researchers at the University of London, England.

In 2014, Green Robot Machinery Ltd. in Bangalore, India was founded by Manohar Sambandam and Neeraj Bhardwaj. When Sambandam chose to turn entrepreneur at the age of 50, he had it all figured out, with a Masters in Electronics & Communication from the Indian Institute of Science, and having worked for 25 years in the semi-conductor industry at Texas Instruments, Athena and Broadcom. He also owns 5 hectares (12 acres) of land and cultivates paddy, cotton and legumes. He is an active participant in various cotton farmers' guilds and is sought after for his views and advises how to leverage technology in farming. He teamed up with his friend and former colleague at Texas Instruments, Neeraj Bhardwaj, to get around the problem of how to disseminate his know-how. Their GRoboMac harvests primarily cotton but also horticulture crops such as egg plant, okra and capsicum. The working of the robot has been designed in such a way that its computerised vision detects and locates the precise 3D coordinates of the bloomed cotton from the images of the cotton plant. A robotic arm uses these coordinates to pick the cotton, and the arm then uses a vacuum for precision picking of the cotton and avoids picking any other contaminant. In 2016, Green Robot Machinery was awarded the Qualcomm Design in India prize. The cotton-harvesting machine has been field tested and became available commercially from June 2019.

Since 2016, Adi Nir and Omer Nir, founders of MetoMotion in the Jezreel Valley Area, Israel have been developing GRoW (Greenhouse Robotic Worker) for picking tomatoes. With a B.Sc. and a M.Sc. in mechanical engineering from the Technion-Israel Institute of Technology, Adi Nir has over 15 years' experience in multidisciplinary product development in the aerospace industry, including management, system engineering, and project management. With a B.Sc. in mechanical engineering from Ort Braude Engineering College and an M.Sc. in agricultural engineering also from the Technion-Israel Institute of Technology, Omer Nir's experience is with advanced product development of robotic systems for agricultural applications. Currently the patent is pending for their proprietary end-effector. The 3D vision system detects ripe fruit and calculates their location. The system is designed to locate a stem without the need for exact data and then to cut and catch in a single operation. It will clear away obstacles without damageto either plant nor fruit.

Farm Bot is an open source precision agriculture CNC (computer numerical control) farming project consisting of a Cartesian coordinate robot farming machine, software and documentation including a farming data repository. The FarmBot project was started in 2011 by American Rory Aronson whilst studying mechanical engineering at California Polytechnic State University at Saint Luis, Obispo. Aronson attended an elective course in organic agriculture where he learned about a tractor that used machine vision to detect and cover weeds which removed the need for herbicides or manual labour: the tractor cost over $1 million USD. In March, 2014 Aronson began working on the project full-time funded by a grant from the Shuttleworth Foundation, to provide hardware kits and software services and to serve as a funding source to maintain the open source community.

The commercial product FarmBot *Genesis*, a small-scale robotconstructed from aluminium, is driven by 17 electric motors, an Arduino and a Raspberry Pi 2 host computer. It can vary in size from a planting area as little as 1m^2 to greater than 50m^2 (11ft^2 – 538 ft^2) while accommodating a maximum plant height of about 1m (3.3 ft). With modifications to some of the structural component sizes and an alternative X-direction drive system, the Genesis concept could likely scale up to a 1000m^2(10,764 ft^2) planting area and a maximum plant height greater than 2m (6.6 ft). The V0.9 seed injector works by using a strong vacuum pump to suction-hold a single seed onto the plastic tip. It is made of a single 3D printed plastic piece with a few other components screwed onto it for magnetic coupling and electronic verification with the UTM.

By late 2018, FarmBot had shipped out hundreds of pre-orders from their third and largest manufacturing run since launching their first product in 2016. FarmBot customers are split 50/50 between the United States and overseas. Nearly half of FarmBot's customers come from the education sector from kindergarten through grade twelve to universities using Farmbot as a research tool, for example Ohio State University which in 2019 received a NASA X-Hab Innovation Challenge award for their Farmbot-integrated project for sustainable food production in space. The other half are "geek gardeners" and commercial agricultural companies A Farmbot currently operating on the rooftop of iTelligence in Berlin grows strawberries and healthy-looking greens. The Berlin Farmbot plot is run by Dries Guth, a Principal Lead at IoT Innovation Lab.[1]

In Italy from 2017, Andrea Gasparri at the University of Rome "La Sapienza" has been leading the H2020 project *PANtHEOn* to design an integrated system where a limited number of unmanned aerial and ground robots move within the hazelnut orchard to collect data and perform typical farming operations. The data is stored in a central operative unit where the information is integrated in order to perform automatic feedback actions, for example to regulate the irrigation system, and to support the decisions of the agronomists. The proposed SCADA (Supervisory Control and Data Acquisition) system will be experimentally validated in a real-world hazelnut orchard located in the Tuscia area in the province of Viterbo, Italy.

In 1998, Nicola Gallo and Alberto and Tiziano Piva founded Ortomec (orto = vegetable garden in Italian) in Cona, Venice region, Italy to develop harvesting and seeding machines. After a few years of tests, studies and development, by 2016 they had produced *Model 4000* for harvesting lettuce heads. It is able to harvest the product by loading it on the machine, keeping it in the same position that it was in the soil without touching it. This method avoids dirtying and/or damaging the plant. The following year came the *Model 9700 Lattuga* enabling the harvest of several kinds of head lettuces, such as Romaine, Butterhead, Batavian. baby leaf, rocket, parsley, spinach, corn salad and beet. The cutting height is adjusted from the driver's dashboard: a swinging knife cuts the heads at the base. Additionally, electronic sensors are in use for automatic and precise cutting height adjustment as the planting bed top may change. The conveyor belt moves the lettuce over the main platform, where the operators can trim and clean the heads. Next, the heads are moved to the rear of the harvester where the crew tidily puts them inside the cartons. In 2019, Ortomec produced the*Herbex II* electrical harvester for Lamb's lettuce. The new system harvests plants via metal shafts that lift, clean and take the produce from the soil to the conveyor. The machine is available in the electrical or combustion engine versions as well as with tracks or wheels, so it can be used either in open fields or in greenhouses.

Lee Redden is well qualified to develop a robotic crop maintenance machine. An engineering intern at NASA, Redden progressed to team leader at the Robotics Academy of Johns Hopkins University Applied Physical Laboratory, then became National Science Foundation (NSF) Graduate Fellow at Stanford University, specialising in computer vision and machine learning applied to robotics. In 2011, Redden started up Blue River Technology in Palo Alto, to develop robots that use computer vision to recognize and identify the needs of plants and take appropriate action in real time, even in rugged agricultural conditions. For example, the robots can determine whether a crop needs thinning, then spray a killing agent with precision to rid excess plants and thin the field.*LettuceBot 1* is a BRT machine-learning powered machine that is able to photograph 5,000 plants a minute, using algorithms and machine vision to identify each sprout as lettuce or a weed. The plants can be identified by a graphics chip in just .02 seconds. *LettuceBot* is also able to determine whether crops have been planted too close to each other, which could inhibit their growth. If that is the case, it will spray and kill one of the plants without harming the other, increasing overall crop yield. This automates a normally labour-intensive process. With *LettuceBot 1* currently in use on 10% of the lettuce fields in the U.S., *LettuceBot 2* was able to perform the real-time thinning of crops four times faster than its predecessor.

Martin F. Stoelen is leading the ABC (Automated Brassica harvest in Cornwall) Project for automating manual picking operations in the horticultural sector in Cornwall and the Isles of Scilly. The main crops for the project will be cauliflower, broccoli, kale and cabbage which are extensively grown in Cornwall. Between 2005 and 2008 Stoelen obtained his M.Sc. in aerospace engineering at the University of Maryland, USA. He then moved to the Carlos III University in Madrid, Spain where in 2014 he obtained his PhD in electrical engineering, electronics and automation ("Sobresaliente Cum Laudem"). He then set up a Soft and Adaptive Robotics (SAR) lab at the University of Plymouth, where his team developed soft (variable-stiffness) robot arm technology, and work to apply it to real-world problems, such as selective harvesting tasks in horticulture. The current projects in the SAR lab include research into autonomous and selective harvesting of raspberries and cauliflower in the U.K. and tomatoes in the U.K. and China. Cameras and sensors in its dexterous fingers can also assess the crop as it grows, so it can determine exactly which vegetables to harvest and which to leave, as a human farmer would know instinctively. In 2018, a test robot was deployed in a cauliflower field near Hayle, Cornwall, owned by David Simmons of Riviera Produce.While Simmons has been working in the industry for 30 years, his family has farmed at Hayle in Cornwall since the 1870s. To commercialise the robotic arm, in 2016 Martin Stoelen founded Fieldwork Robotics Ltd.

Established in 2015 in Qadima Zoran, Israel, FFRobotics, has been testing its technology in apple orchards in Washington. Using an innovation made by Avi Kahani for Ffmh-Tech Ltd., multi-robot crop harvesting machine involves robots mounted on a common frame facing an area to be harvested, and are each configured to harvest crop items by reaching and gripping the crop items from a fixed angle of approach. The sensors are configured to acquire images of the area and the computers are configured to identify the crop items in the images, and to direct the robots to harvest the identified crop items. The FFRobot can thus precisely and gently pick ten times more usable fruit compared with an average worker.

Vitibot is the creation of Cédric Bache, the son of a family of winemakers in the Champagne region who has always been immersed in robotics. During his studies at UTC Compiègne, Bache began to study the way to develop an autonomous grape-harvesting robot to replace the classic straddle tractor in the vineyards. His first prototype *Hector* (as in hectare) began trials all over the Champagne wine-growing region in May, 2015. Three prototypes and fifteen months later, *Bakus* (as in Bacchus, the Roman god of wine) 100% autonomous and 100% electric, capable of working for up to 10 hours, equipped with 8 cameras was ready. Early in 2018, having won the Innovation Trophies competition at the ViniTech Sifel Show, Vitibot moved to 4000m^2 (43,000 ft^2) industrial premises Reims for further development and field trials.

J.F.M. Huijsmans and a team at Wageningen University, Netherlands have developed an autonomous orchard robot, using LiDAR and V-shaped turning. In mountainous orchards, agricultural tasks such as crop protection and harvesting, are characterized as being labour-intensive and dangerous. An autonomous orchard robot that can execute these unattended seems a promising alternative to increase task operability. This is achieved using a probabilistic particle filter (PF) algorithm, with a novel laser-beam model for autonomous in-row navigation.

Brandon Alexander has created Iron Ox, a fully autonomous farm in San Carlos, California. Alexander had previously worked at robotics lab Willow Garage and later at Google parent company Alphabet's secretive X division (the R&D lab behind Google Glass, self-driving car spinoff Waymo, and Loon). He and cofounder Jon Binney, Iron Ox's CTO, set out to drive down produce prices with the goal of helping sustainably feed 10 billion people. Their hydroponic indoor farm relies on two robots to plant, care for and harvest produce. One of the robots is 454 kg (1,000 pounds) and about the size of a car. It picks up the trays of plants and transports them around the greenhouse. A second machine, a robotic arm, is responsible for all the fine manipulation tasks, such as seeding and transplanting. As a tray of plants matures, the mobile robot carries it to the processing area. Here, the robotic arm moves baby plants in densely packed trays to containers with more space. This optimizes space efficiency, because throughout their life cycle, plants are only given the room they need. Alexander has claimed that Iron Ox is able to do the equivalent of 12 hectares (30 acres) of outdoor farming in just less than half a hectare (1 acre) on its robotic farm.

Iron Ox wants to build more small farms near urban centres so produce is fresher upon arrival. The robots at Iron Ox also use machine learning and AI to detect pests and diseases. They can remove infected plants before the problem spreads. Though Iron Ox grows its produce using LED lights, in the future it hopes to build natural-light greenhouses to take advantage of the sun's free energy. Eventually, the company aims to make its non-GMO and pesticide-free produce as cheap as traditional agriculture. The farm is currently growing a number of leafy greens, including Romaine, Butterhead and kale, in addition to basil, cilantro (coriander) and chives. The robots tending these plants are *Angus*, a 454 kg (1,000 lb) machine that can lift and move the large hydroponic boxes in which the produce is growing, and Iron Ox's robotic arm for harvesting the produce. From April, 2019 Iron Ox began to offer three varieties of greens at the San Carlos branch of Bianchini's Market, a family-owned grocery store that specializes in local and organic produce.

In 2019, the world's first Centre for Doctoral Training (CDT) in agricultural automation and robotics was opened in Lincolnshire. Professor Tom Duckett, Professor of Robotics and Autonomous Systems at Lincoln University, was the new Centre Director in collaboration with the University of Cambridge and the University of East Anglia. The new training centre in agri-food robotics will create the largest ever cohort of Robotics and Autonomous Systems (RAS) specialists for the global food and farming sectors, thanks to a £6.6 million funding award. The British CDT will provide funding and training for at least 50 doctoral students, who will be supported by major industry partners and specialise in areas such as autonomous mobility in challenging environments, the harvesting of agricultural crops, soft robotics for handling delicate food products and 'co-bots' for maintaining safe human-robot collaboration and interaction in farms and factories. Agri-food is the largest manufacturing sector in the U.K. – twice the scale of automotive and aerospace combined – supporting a food chain, from farm to fork, which generates a Global Value Added (GVA) of £108bn, with 3.9m employees in a truly international industry.

Autonomous robotic agriculture is not only about plants

Lely Holding in the Netherlands manufacture cow milking robots. Wide Valley Farm, a small dairy operation in Pennsylvania's Centre County deploysone of these. Their 60 cows, lured by special feed, enter a stall one by one throughout the day to be milked by a robotic arm. Around each cow's neck is a bovine Fitbit of sorts, an activity sensor that gives a constant readout of her movements, chewing, daily milk production and other health indicators. But processing data from the sensors and the Lely milking robot is a challenge. Commercial telecoms have not delivered broadband to their sparsely populated valley. So, the dairy uses a "hotspot" device that relies on weak cellular signals to get that data to their computers. They use it 24 hours a day. Milking robots are one of the most prominent robots in the world, with a market expected to double to 2.48 million by 2023.

In 1995 Lely introduced their Astronaut daily robotic milk system. By 2014 the 20,000th Lely Astronaut had been installed. (Lely)

In June, 2019 a Dutch property company called Beladon launched the world's first floating dairy farm anchored to the ocean's floor in the middle of Rotterdam's Merwehaven harbour. 40 Meuse-Rhine-Issel cows are milked by Lely robots to produce 800 litres (176 gallons) of milk a day. Peter van Wingerden, an engineer at Beladon, came up with the idea in 2012 when he was in New York working on a floating housing project on the Hudson River. While there Hurricane Sandy struck, flooding the city streets and crippling its transport networks. Deliveries struggled to get through and within two days it was hard to find fresh produce in shops. Seeing the devastation caused by Hurricane Sandy van Wingerden was struck by the need for food to be produced as near as possible to consumers.

The top floor of the cow garden houses greenhouses for clover, grass and other crops that feed the cows; the middle level is the animal's floating home, a grassy enclosure meant to resemble a natural garden, but populated with artificial trees. The cows are free to roam in and out of their stalls, and also have the option to graze on solid ground in an adjacent field they can access via a ramp.

Lely robots provide them with fodder and brush them down. Finally, the bottom floor contains a processing plant, turning fresh milk into consumer products, including yogurt and, possibly, Comté-style cheese. All of this futuristic food manufacturing his happening behind glass walls, to literally emphasize transparency. School children and consumers are being invited to tour the farm and watch robots milk the cows and pick up their waste, which is used as fertilizer or converted into energy for on-site use. To power it, the farm uses solar and wind energy from rooftop windmills and solar panels, while artificial trees with real ivy provide shade for the cows and reduce energy consumption by cooling the space. The cows came on board in April, 2019.One of the Dutch farming organizations that collaborated with Beladon is planning to take the same concept to other cities, and is already developing a floating egg farm. Before they move on to chickens however, they'll hopefully have solved one current problem: what to do about that distinct animal farm odour.

In 2007, Bastiaan Vroegindeweij and three fellow students at Wageningen University in the Netherlands entered an international robot competition Field Robot Event with a prototype with a "pick-up cart" behind it for the eggs lying on the ground, manually controlled via a laptop. In an article in the trade journal *Poultry Farming* in August, 2007 Vroegindeweij said that a final design would take another ten years, and also that "Perhaps the subject of computers and robots in poultry farming will become my graduation assignment and I will continue to do so even after graduation." And that happened, first as a student at Wageningen UR, then for five and a half years as a researcher and since then with his own company, Livestock Robotics.

Chickens lay most of their eggs in their nests, but some end up in the hay. Egg collection is both necessary and time-consuming. *PoultryBot* can move through the stall freely. The robot identifies and collects missing eggs, thereby saving poultry farmers thousands of euros per year. The robot is also equipped with a device that can measure the stall climate or look after sick birds. The chickens adapted to the new addition rather quickly and soon saw it as a playmate. With this robotic egg collector, Wageningen UR is working on improving quality of life. *(See also Chapter 2 for the history of Battery Brooding in the 1930s).*

In Cesson-Sévigné, Brittany, France since 2016, Yann Courcoux and Pierre Mabit, a Mecatronics engineer, have developed the Tibot Spoutnic or *TiOne*, an autonomous poultry robot which encourages errant hens to lay eggs in their nests. It circulates in poultry houses 24/24 hours and solves the problems of laying on the ground. Untrained chickens may lay eggs wherever they see fit, leaving farm workers to collect, which poses both sanitary issues as well as inefficient production. The robotic chicken coop worker operates by scanning the barn, prompting the animals to move around and lay eggs in nests, rather than on the floor - real sports coach. In the end, there is a decline in the mortality rate on farms and an improvement in working conditions.

Ian Hunter, Professor of Mechanical Engineering at MIT is the co-founder of 25 companies, and has more than 100 patents concerning micro-instrumentation, micro-fabrication, micro-robotics and medical devices such as needle-free injection to his name. Fonterra Co-operative Group Limited is a New Zealand multinational dairy co-operative owned by around 10,500 New Zealand farmers. The company is responsible for approximately 30% of the world's dairy exports and with revenue exceeding NZ$17.2 billion, is New Zealand's largest company. In 2018, Fonterra visited Hunter at one of his companies, Indigo, which designs electric vehicles. The result was an agrobot with a multitude of uses, including collecting cow urine and excrement to reduce effluent run-off, also deploying the needle-free drug delivery technology. After further tweaks and testing, Ian Hunter, with his brother Peter Hunter, the Director of the Auckland Bioengineering Institute, hope small fleets could one day potter about farm pastures, gathering excrement, neutralising urine patches and delivering precision doses of fertiliser directly to grass roots, after first conducting soil tests to see what nutrients are missing. The Hunter agrobots would act as guardians of the herd, moving around in the paddocks picking up cow dung, moving it to a bioreactor in one corner of the paddock from which electrical energy could be used to run all of the farm equipment. Biogas from the manure, combined with solar energy, could charge the robots twice daily via a wireless flash charge. The clean energy could also dry any milk that was going to be trucked off and sold as milk powder. Drying milk on the spot would remove the need for transporting heavy liquid milk by road for drying by coal-fired boilers, saving energy and fuel. The water extracted from liquid milk could be re-used on the farm for irrigation. The Hunter brothers and Fonterra are not imagining one robot per farm. If each cow produces 30kg (66 lb) of dung a day, 50 cows per robot might be manageable, making a fleet of six robots for 300 cows, or 10 robots for a farm with 500 head of cattle. As well as being cheap, they would need to be non-threatening to cattle, to avoid getting a hoof in the motor. Hunter hopes they would be non-threatening to farmers, too. "We're hoping to make life much more pleasant for the farmer," he says. "The last thing they want to be doing is rushing around spraying urine to neutralise it, or picking up cow patties.[2]"

Can a drone and a bot be combined into one machine in the service of agriculture? In 2019, Professor David Zarrouk, head of Ben-Gurion University of the Negev (BGU)'s Bio-Inspired and Medical Robotics Lab, and his graduate student Nir Meiri demonstrated their hybrid FSTAR (flying sprawl-tuned autonomous robot). This robot drone flies like a typical quadcopter, drives on tough terrain and squeezes into tight spaces using the same motors. It also adjusts its width to crawl or run on flat surfaces, climb over large obstacles and up closely spaced walls, or squeeze through a tunnel, pipe or narrow gaps.

Cyborg Botany is a new, convergent view of interaction design in nature which may herald a new form of agriculture. In 2018, Harpreet Sareen and Pattie Maes, working at the Massachusetts Institute of Technology's Media Lab's Fluid Interfaces group have developed *Elowan*, a cybernetic life form or a robot plant-electronic hybrid.[3] *Elowan* has a wheeled robot attached to its base and electrodes embedded in its leaves and stems. Its electrodes pick up weak bioelectrical signals the plant produces naturally in response to light and other environmental changes. These signals trigger the robot to move, bringing it closer to or further away from a light source. The researchers state:

> In recent history, humans domesticated certain plants, selecting the desired species based on specific traits. A few became house plants, while others were made fit for agricultural practice. From natural habitats to micro-climates, the environments for these plants have significantly altered. As humans, we rely on technological augmentations to tune our fitness to the environment. However, the acceleration of evolution through technology needs to move from a human-centric to a holistic, nature-centric view.

Should one be wary of all these *bots?* Some international financial institutions such as the World Bank, OECD (Organisation for Economic Co-operation and Development) and Oxford University have suggested that automation could wipe out almost half of all jobs in twenty years. Yet the future of the job sector relies heavily on developing social skills, trust and empathy among workers.Consider the longevity we are enjoying nowadays and the difficulties of the education sector in many countries and the problem is compounded. One proposed solution is a universal basic income doled out by the government, a sort of baseline one would receive for survival. After that, re-education programmes could help people find new pursuits. Others would want to start businesses or take part in creative enterprises. It could even be a time of the flowering of humanity, when instead of chasing the almighty dollar, people would able to pursue their true passions.

Last but not least: Vertical Farming

Vertical farms depend on multiple arrays of light emitting diodes.

The electrical agricultural machinery used for vertical farming includes diode lights and watering systems. Vertical farmers tend to choose from hydroponics in which plants are grown in a nutrient-rich basin of water, and aeroponics, where crops' roots are periodically sprayed with a mist containing water and nutrients.

Perhaps the earliest example of a "vertical farm" is the legendary Hanging Gardens of Babylon, built by King Nebuchadnezzar II more than 2,500 years ago. According to some scholars, the gardens consisted of a series of vaulted terraces, stacked one on top of the other and planted with many different types of trees and flowers. Reaching a height of 20 metres (66 ft), the gardens were likely irrigated by an early engineering innovation known as a chain pump, which would have used a system of buckets and pulleys to bring water from the Euphrates River at the foot of the gardens to a pool at the top.

One of the earliest drawings of a tall building that cultivates food was published in *Life Magazine* in 1909. The reproduced drawings feature vertically stacked homesteads set amidst a farming landscape. Six years later, the term "vertical farming" was coined by American geologist Gilbert Ellis Bailey working for DuPont de Nemours in his book of the same name (Lord Baltimore Press). Interestingly, Bailey focused primarily on farming "down" rather than "up." That is, he explored a type of underground farming in which farmers use explosives to be able to farm deeper, thus increasing their total available area and allowing for larger crops to be grown.

In his article "Aquaculture: A means of Crop-production," published in December, 1929 William F. Gericke, an agronomist at the University of California, Berkeley, outlines the process of growing plants in sand, gravel or liquid, using added nutrients but no soil. The term "hydroponics" was coined in an article published in 1937 in *Science* magazine. Derived from the Greek words "hydro" (water) and "ponos" (labour), the term was suggested to Gericke by his University of California associate, botanist William Albert Setchell as an alternative to "aquaculture" which was already in use to describe fish-breeding techniques.

Images of the vertical farms at the School of Gardeners in Langenlois, Austria, and the glass tower at the Vienna International Horticulture Exhibition (1964) show that vertical farms existed. The Armenian tower hydroponicums are the first built examples of a vertical farm, and are documented in Sholto Douglas' "Hydroponics: The Bengal System", first published in 1951 with data from the Bangladesh (then known as East Pakistan), and the Indian state of West Bengal.

In 1999, the concept of the modern vertical farm was developed in a class led by Columbia University environmental health sciences professor Dr Dickson Despommier. In an effort to figure out an effective way to feed the population of New York using only urban rooftop agriculture, Despommier and his students developed the idea of a 30-story urban farm with a greenhouse on every floor: in other words, a contemporary vertical farming tower. (Despommier has since gone on to become the world's foremost expert on and proponent of vertical farms.)[4]

Since the late 1880s, when Carl William Siemens had carried out plant-growth experiments with primitive electric lighting in his greenhouse at Sherwood, Tunbridge Wells, England, electric lights have been used to speed up plant growth. From the early 1970s, the light-emitting diode (LED), a semiconductor light source that emits light when current flows through it, was seen as a more efficient replacement for fluorescent light bulbs for plant growth. Compared with other forms of electrical illumination, light-emitting diodes use less energy, give off little heat and can be manipulated to optimize growth. In 2009, Siddha Pimputkar, a postdoctoral researcher in the Materials Department of the Solid State Lighting and Energy Electronics Centre, the University of California, Santa Barbara published an article "Nature Photonics: Prospects for LED Lighting" in which he pointed out the advantages of diode lamps for plant growth. As lighting supplement, LEDs can be used in anytime and anywhere: no matter in daylight or night, crops can bath in enough light emitted by LED. On one hand, people can never savour the taste of the fruit impacted by the shortage of sunlight. On the other hand, LEDs enable us to plant all kinds of crops throughout every season. In short, they would be perfect for vertical farming. In 2006,

Nuvege (pronounced "new veggie") of Kyoto, Japan developed one of the essential ingredients for indoor vertical farms: a proprietary LED network that balances light emissions in order to increase the return rate of vegetables. Its LED lighting was tuned to service two types of chlorophyll, one preferring red light and the other blue. Nuvege produces 6 million lettuce heads a year.

In the 1970s, Ken Yeang, a Malaysian architect, began to propose that instead of hermetically sealed mass-produced agriculture, plant life should be cultivated within open air, mixed-use skyscrapers or Farmscrapers for climate control and consumption. This version of vertical farming Veg.itecture, is based upon personal or community use rather than the wholesale production and distribution that aspire to feed an entire city.

In 2006, Shinji Inada, a former vegetable trader, founded SPREAD, and opened his first automated indoor vertical farm facility the following year in Kameoka, Japan. The company spent years refining systems for lighting, water supply, nutrients and other costs. By 2013, producing 21,000 heads of lettuce per day SPREAD finally turned its first profits, and at the time made it the world's largest vertical farm in terms of production. Their brand, "Vegetus" was soon available in approximately 2,400 supermarkets nationwide. In 2014, a partnership system and global expansion strategy for SPREAD's vertical farming business was established. In 2015, they announced the concept for the leading vegetable production system. This new low cost and more environmentally friendly system would first be constructed in a new facility called Techno Farm Keihanna in the Kansai Science City, designed as the world's largest automated leaf-vegetable factory with an output of 30,000 heads of lettuce a day or 648 heads of lettuce a square metre (11 ft^2) annually, on racks under custom-designed lights using LED. A sealed room protects the vegetables from pests, diseases and dirt. Temperature and humidity are optimized to speed growth of the greens, which are fed, tended and harvested by robots. The Techno Farm will use only 110 millilitres(4 fl oz) of water a lettuce, 1 percent of the volume needed outdoors, as moisture emitted by the vegetable is condensed and reused. Electrical power consumption per head will also decrease with the new factory using custom-designed LEDs that require about 30 percent less energy. A collaboration with telecoms company NTT West on an artificial intelligence program to analyse production data could boost yields even more. Shinji Inada won the Edison Award in 2016 for his vertical-farming system.[5]

SPREAD is not alone. As of 2014, Vertical Fresh Farms was operating in Buffalo, New York specializing in salad greens, herbs and sprouts. In March, the world's then largest vertical farm opened in Scranton, Pennsylvania, built by Green Spirit Farms (GSF). The farm is housed in a single-story building covering 3.25 hectares (8 acres), with racks stacked six high to house 17 million plants. The farm grows 14 lettuce crops per year, as well as spinach, kale, tomatoes, peppers, basil and strawberries. Water is scavenged from the farm's atmosphere with a dehumidifier.

JX ANCI, a wholly owned subsidiary of JXTG Nippon Oil & Energy Corp., plans to build an indoor farm in its Narita plant by 2020, using SPREAD's system. And Mitsubishi Gas Chemical Co. has agreed with Farmship Inc., a Tokyo-based start-up established by a former SPREAD employee, to build an indoor farm in Fukushima Prefecture that would grow 32,000 lettuces a day.

From 2008, Glen Kertz of Valcent, a tech company based in El Paso, Texas developed hydroponic greenhouse methods to grow upward rather than out. Potted crops grow in rows on clear vertical panels that rotate on a conveyor belt. Moving them gives the plants the precise amount of light and nutrients needed, an optimization that enables Kertz to grow 15 times as much lettuce per acre as on a normal farm, using 5% of the water that conventional agriculture does. Valcet's "Vertigro" closed vertical growth system now produces over a million litres per hectare of algae biofuel each year, (100,000 gallons/year/acre).

AeroFarms in Newark, New Jersey a sustainable 6,500 m^2 (70,000 ft^2) vertical farm located in a former steel factory in the Ironbound section, leans heavily on intelligent machines. There, 907,000 kg (2 million lb) of greens and herbs are produced each year, using an aeroponic growing method. Aeroponic growing towers are a closed-loop system, recycling the water and nutrients with virtually zero waste. AeroFarms patented growing system mists the roots of their plants with targeted nutrients, water and oxygen. This system uses up to 95% less water than field farming to grow high-quality produce faster and more efficiently, with zero pesticides. After testing hundreds of growth media for the plants, they have developed a patented, reusable cloth medium made out of 100% recycled materials for seeding, germinating, growing and harvesting. With projects in development in China, the United Arab Emirates and Europe, AeroFarms has its sights on the world, but is still very focused on Newark where they have four farming operations and employ over 120 people, 40 percent of whom live in Newark, with 80 percent within a 15-mile radius. AeroFarms produce is sold through their retail brand, Dream Greens. Meanwhile, Bowery, which is growing crops inside two warehouses in New Jersey, can promise people in New York that their "bok choy" did not travel far at all.

Some commercial ventures have targeted wealthy nations in the Middle East as prime candidates for vertical farms because of the high cost of importing fresh produce. In 2019, Dubai's Emirates Flight Catering began construction of a 12,000 m^2 (130,000 ft^2) vertical farm, located near Al Maktoum International Airport at Dubai World Central to supply airlines in a joint venture with California-based Crop One Holdings. The $40 million facility is expected to deliver its first vegetables to airlines and airport lounges in December, 2019. At full productivity the project will produce 2,700kg (6,000 lb) of herbicide-free and pesticide-free leafy greens every day.

Other high-rise farms have appeared in office towers or condos as part of the design. In Tokyo's Ginza shopping area, stationery retailer Itoya tends a vertical farm on the 11th floor of its 12-story building to supply lettuces exclusively to its cafe, at a cost that would be uncompetitive with vegetables grown in outdoor farms.

Berlin-based in Farm has developed a vertical indoor farming system that can be implemented in supermarkets, restaurants, local distribution warehouses, or even schools, allowing businesses to grow their own fresh produce on site to deliver to customers. InFarm controls the farms remotely using IoT, Big Data, and Cloud analytics. Founded in 2012, the start-up is already opening indoor farms in 1,000 locations in Germany and expanding in other European markets, while planning to launch in North America in 2019.

In early November, 2016 Tesla entrepreneur Elon Musk's brother, Kimball, and fellow entrepreneur Tobias Peggs launched Square Roots, an urban farming incubator programme in Brooklyn, New York. The setup consists of 10 steel shipping container farms with 30 m^2 (320 ft^2) vertical growing space where eight young entrepreneurs, including Josh Aliber and Electra Jarvis, work to develop vertical farming start-ups. The USDA gave the Square Roots entrepreneurs small loans to cover preliminary operating expenses. Other investors include Powerplant Ventures, GroundUp, Lightbank and FoodTech Angels. In 2019, Square Roots announced that it was taking its ultra-scalable urban farming model nationwide with Gordon Food Service.

Other chapters in this book report on the use of electric or solar-electric flying machines to help with precision agriculture: satellites and drones. There is one other electric agricultural machine circulating the Earth: the International Space Station (ISS)

The sky is the limit!

Plants have actually been grown in space for a long time. In 1982, Soviet cosmonauts on board the *Salyut 7* space station nurtured some thale cress, *Arabidopsis thaliana*, making it the first plant to produce flower and seed in space. The International Space Station (ISS) is a space station, or a habitable artificial satellite, in low Earth orbit. Its first component was launched into orbit in 1998, with the first long-term residents arriving in November, 2000. The ISS maintains an orbit with an altitude of between 330 and 435 km (205 and 270 mi) by means of re-boost manoeuvres using the engines of the Zvezda module or visiting spacecraft. Without thermal controls, the temperature of the orbiting ISS's sun-facing side would soar to 250 degrees F (121 C), while thermometers on the dark side would plunge to minus 250 degrees F (-157 C).

Despite these conditions, from 2014 astronauts on the ISS have been growing and eating fresh vegetables in orbit. That year, when SpaceXlaunched its *Dragon* cargo mission from Cape Canaveral to the ISS, it took the Veg-01 minifarm, nicknamed "Veggie" with it. The astro-plants in Veggie included lettuce, Swiss chard, radishes, Chinese cabbage and peas. Red Romaine lettuce grown in space on Expedition 40, were harvested when mature, frozen and tested back on Earth. Expedition 44 members became the first American astronauts to eat plants grown in space on 10th August, 2015 when their crop of Red Romaine was harvested.

In 2017, the Advanced Plant Habitat was designed for ISS, installed in parallel with Veggie. The major difference with that system is that APH is designed to need less upkeep by humans. In December, 2017 hundreds of seeds including dwarf wheat and *Arabidopsis* were delivered to ISS for growth in the Veggie system. In 2018, the Veggie-3 experiment was tested with plant pillows and root mats. Crops tested at this time include cabbage, lettuce, and mizuna. In 2018, the PONDS (Passive Orbital *Nutrient Delivery System*) system for nutrient delivery in microgravity was tested. In December, 2018 the German Aerospace Centre launched the EuCROPIS satellite into low Earth orbit. This mission carries two greenhouses intended to grow tomatoes under simulated gravities of the Moon and Mars using by-products of human presence in space as source of nutrients.

In his plan to use SpaceX starships to colonise the planet Mars before 2030, Elon Musk envisages the first settlement using hydroponics for growing plants. In 2019, Musk told *Popular Mechanics* magazine:

You essentially have solar power — unfoiled solar panels on the ground, feed that to underground hydroponics, either underground or shielded by wires, dirt, so then you can be sure that you don't have to worry about excessive ultraviolet radiation or a solar storm. Really pretty straightforward. You could just use Earth hydroponics. Earth hydroponics will work fine.[6]

Hydroponics also plays a role in science fiction. In the "Star Trek" films and television series (2001-2005). Starship *Enterprise NX-01* launched by the United Earth Starfleet in 2151, maintains a hydroponic greenhouse to provide fresh fruits and vegetables for the crew's mess hall. In this prequel series, *Star Trek: Enterprise* set in the 22nd century, the ship features a "protein resequencer" that can only replicate certain foods, so an actual chef serves on board who uses the hydroponic greenhouse for his ingredients. Time will tell.

Humans are still cheaper than robots, and so they are at present used for the vast majority of farming. Robot farming on land or in the air might just be a niche. But it is an important niche.

Endnotes

[1]Elizabeth Rushe, "Hundreds of California Garden Robots Migrate Around The World," Forbes, September 18th 2018.

[2] Eloise Gibson, "Celebrated brothers turn brains to dairy problems," newsroom.co.nz, May 30, 2019.

[3]In "Starflight, a 1980s space exploration, combat, and trading role-playing video game, the Elowan are a pacifistic race of plant creatures native to the Alpha sector.

[4]Mark Crumpacker, "A Look at the History of Vertical Farming," Medium.com, October 19, 2018.

[5]Aya Takada, "As high-rise farms go global, Japan's Spread leads the way," The Japan Times, November 1st 2018.

[6]Ryan d'Agostino, "Elon Musk: The Popular Mechanics Interview" February 25, 2019.

Appendix A

e-weed killers

George Washington Carver, an American agricultural scientist and inventor once said "A weed is a flower growing in the wrong place." Nevertheless they can be a menace to agriculture.

In 1895, Albert A. Sharp of Memphis, Tennessee invented a new and improved Vegetation Exterminator: "The object of my invention is to produce a simple apparatus which may be conveniently applied to a car or other vehicle, and which when the car to which it is applied is moved along the track, or a vehicle is moved along the road, will cause a strong current of electricity to be sent through all the adjacent vegetation, thus killing the same" (Patent US546682A). Sharp's idea was to kill the rank vegetation which grows along railroad banks and highways in tropical countries.

Through the years, a handful of inventors would tackle the challenge of the electric weed killer. One of these was William E. Burt of the small town of Yuba, Wisconsin who had invented a dynamometer for measuring horsepower back in 1909. In 1928 he obtained a patent for destroying weeds, whereby an oscillatory current would pass through the root system of a plant such as Canadian thistles which were very hard to destroy by known methods. Burt suggested a powerful current transported from place to place by an electric wagon or truck. On board was a motor, a DC generator or exciter and an AC generator. The farmer would then use a switch to oscillate the current and killed the weeds.

In 1947, Russell R. Poynor of Elmhurst, Illinois, assignor to the International Harvester Corporation, applied for a patent for an electric weed killer, "relating to the destruction of weeds and the like which tend to interfere with and retard the growth of selected vegetation. The invention concerns particularly a machine for thecultivation or crop plants and apparatus therefore. A further object of the invention, therefore, is the provision of mechanism mounted upon a power propelled vehicle carrying a member of electrically conductivematerial having sufficient rigidity to brush over weeds growing in a crop row but being flexibly mounted to permit the member to yield upon encountering the greater resistance of the crop plant."(US 7933.43)

Enter Lascelles A. Chin,one of Jamaica's most distinguished and respected entrepreneurs, a pioneering exporter, outstanding philanthropist, and a much honoured leader in Jamaica and the Caribbean. In the 1970s the Lasco Corporation started to manufacture and sell electrical weed control equipment called the Lightening Weeder. It was patented in 1977 by Ricks H. Pluenneke and Willis G. Dykes as an "apparatus for using electrical current to destroy tall weeds in and around crop rows" which comprised a 125 hp tractor-linkage-mounted machine with a 7 metre (23 foot) swath.When a farmer was not controlling weeds with his Lightning Weeder, he could use it as a dependable standby source of electric power. It produced both 120V and 240V single-phase power to run most electrical equipment such as a welder, for repairing in the field.Several safety interlocks incorporated into the high voltage machine must all be satisfied to make it operate. For example, the machine would not run if the operator was not in the seat, if the machine lost its electrical ground connection, or its forward motion. These patented safety features, along with operator training and supervision, helpedensure safety for both operator and bystanders. Lasco's claimed that they had never heard of a bad experience with electric shock, nor had they had a problem with dry crops catching fire in the field.

They marketed it as their *LW5* Lightning Weeder,built in Vicksburg, Mississippi, and first appearing on the market in 1980, selling at home and abroad to the USSR and to the U.K. U.K. sugar beet growers were particularly interested in the machine as a means of destroying weed beet in their crops. However, this interest and Lasco's machine disappeared because the technique did not achieve commercial success, mainly due to herbicides being better understood by farmers that electrothermal machines, in what was then a small market with the advent of relatively low cost weed wipers which could do the same job at least as effectively.

In 1996, a team of Swedish engineers, Bertil Persson, Pär Henriksson, Tomas Nybrant and Berit Mattsson of Zero Weed AB developed a system where weeds were controlled by high voltage pulses with short duration which electropermeabilized the cell membranes of weed seeds in the ground. The device was selective and damagedonly germinating weed seeds and plants early in their life cycle. The required amount of energy was small; with rectangular pulses the optimal field strength was between 100-300 kV/m with a duration of 10-100 microseconds. A transformer placed on a sowing machine transformed electrical energy to high voltage pulses. The energy could be taken from the pulling tractor via a transmission or from an integrated power source. The high voltage pulses were applied to electrically conductors via applicators through two or more fixed, spaced plates to the soil around newly sown seeds.

The bad publicity concerning the harmful effects of the blanket-spraying of herbicides on the health of farm workers, and consumers' mistrust of genetically modified crops has led to a renewed quest for an efficient electric weed-killer, using the compact, high-energy lithium battery available since 2007.

In 2010, Sergio de Andrade Coutinho set up Zasso ("weed" in Japanese) in São Paulo, Brazilto design and develop an electric weed-killer whose "residue-free electron" system was trialled in the Latin American city then patented by Coutinho with Matthias Eberius and Dirk Vandenhirtzin 2013. Three years later Zasso Digital Herbicide was offering the *Electroherb* whose flexible, interchangeable applicators allow the XPower to be used in a range of surfaces and markets. Units can be adapted for different row crop systems, and can be controlled via ISOBUS Class 3 compatible tractors. In 2017, Zasso developed a self-driving robot system which can continuously measure the soil resistance. During 2018 Zasso built a new R&D centre in Aachen, the leading German city for electronic development. The facility,near to RWTH Aachen Universityis designed to run carbon dioxide neutral and can host nearly 100 engineers on 850 m^2 (9150 ft^2). In addition, Zassohas built a 450 m^2 (4844 ft^2)lab to build and test the latest generation of the digital herbicide platform.

Founded in 2012, Michael Frederik Diprose of Ubiqutek, Kineton, Warwickshire, Englandhas innovatedRootWave,a phalanx of cutting-edge electricide technology. RootWavehas partnered with Steketeeon a pull-behind unit covering 8-12 rows using camera imagery to spot and zap weeds on the go, rolling at roughly 5 km/h (3 mph), with power sourced from the PTO. Essentially, visual recognition identifies weeds in real-time then RootWave makes contact with individual weeds to deliver a 5 kilovolt jolt (5,000 volts) with no soil disturbance.Differences in silt loam, buckshot, heavy sand and general variations in soil types (and moisture content) sometimes require changes in voltage. There are also nuances with root type. Fibrous or taproots arenot an issue, but a rhizome may require multiple passes. RootWave calls this "electride technology".

Over in the USA in 2015,two brothers from Illinois with engineering and farming backgrounds who wanted to tackle both organic weed control and the growing issue of herbicide resistant weeds, established The Weed Zapper, modified and fine-tuned from the technology used in the Lasco Lightning Weeder. Two years later, two more brothers from Sedalia, Missouri, both with organic farming operations of their own, and a commercial hvac (heating, ventilation, and *air conditioning)* and electrical businesses, purchased The Weed Zapper technology and prototype and began production of the new Annihilator series. The three-range Annihilator models are ruggedly-built tractor-powered and attached units that kill weeds, including the root using electricity. Depending on the number of rows to be covered, PTO tractors of from 135–275 hp are used. Annihilatorskill both in-row and between-row weeds, including late-season weeds, in as little as one pass. The unit's electrode bar runs about 10-15 cm (4-6 in) above the crop in soybeans, wheat, rye, milo, peas, sugar beets, vegetables or other legumes or small grains. The units are also highly effective for cleaning up unwanted weeds in pastures. The units are built at theOld School Manufacturing, LLC.

So far, electric weed-killers had been attached to a tractor driven by a farmer.

In 2011 Gaëtan Séverac and Aymeric Barthès started up Naïo Technologies in Escalquens (Haute-Garonne),France to develop an all-terrain weeding robot (Naïo Technologies: Photo Tien Tran).

In France, Naïo Technologies is leading the way. In 2010, Gaëtan Séverac, PhD student in robotics teamed up with Aymeric Barthès, one of his classmates at the Institut Méditerranéend'Etude et Recherche en Informatique et Robotique (IMERIR) to develop an all-terrain weeding robot. *OZ*, their prototype used a satellite positioning algorithm with a precision of 4 cm (1.5 in) called PPP-CNES, (PPP meaning Punctual Positioning Specific). In 2011,Séverac and Barthès founded their start-up, Naïo Technologies inToulouse. Soon after, field trials were carried out on two vegetable farms and a vineyard in the Occitania region. From 2015, Naïo Technologies organised a "Move Your Robot" national contest opened to engineering colleges and universities, with the objective of improving the *OZ* guidance programs. For example, in 2016 participants proposed a power supply solution with a solar panel adaptable to the robot, a touch-screen human-machine interface, a soil analysis laboratory embedded on the robot, a voice guidance system and a gun noise to scare birds. By 2016, a growing number of *OZ* weed-killers were being used by customers anxious to get away from products such as Monsanto's Roundup (glyphosate). Naïo next produced *TED*, a vine-straddling robot weed-killer, trialled by Bernard Magrez up and down the vines of his Château Fombrauge (Saint-Emilion Grand Cru Classé), then by Philippe de Rothschild on his vineyard. Measuring 1.8 m wide by 2m high (6ft by 6.6 ft), equipped with a GPS, the electric 4WD *TED* is able to leave the wine warehouse to go directly to the plot; programmable to work according to the weather, and to make several passes.Naïo Technologies' next machine was *DINO*, a straddle robot for the mechanical weeding of vegetable plantations. It is particularly suitable for salad crops, which it weeds mechanically and autonomously thanks to its hoeing and guiding tools. *BOB* the fourth option runs on caterpillar tracks. In December, 2018 the fourth FIRA International Forum of Agricultural Robotics was held over two days at the Diagora centre in Labège. Organized by Naïo-Technologies, it hosted more than 800 delegates from around the world. This sector is evolving, with projects of all shapes and sizes.

In 2019, Naïo Technologies with a workforce of 50, based in Escalquens (Haute-Garonne), proudly announced the sale of its one hundredth *OZ*.

En 2019, Naïo Technologies proudly announced the sale of its one hundredth OZ. used by customers anxious to get away from products like Monsanto's glyphosate (Naïo Technologies: Photo Tien Tran).

In 2019, Cyril Roudotof Natuition teamed up with Russian Vladimir Modylievskiiwho specialised in electric propulsion,to develop a weed-killer robot called *Violette* at the fablab Les Etablis & Co, in La Rochelle, France.It is equipped with a small camera enabling it to identify the weeds on the ground it is passing over. Photos are then transmitted to a program which compares them to a plant image database, enabling *Violette* to pull out each identified weed by its root. In 2019, 20 *Violettes* went on trial with test-customers, prior to seriesmanufacture in 2020.

The Swiss have developed the ecoRobotix autonomous weeding robot. Steve Tanner grew up on a farm in the Orbe Valley in the Swiss canton of Vaud, and as a child he used to help his parents weed fields of sugar beet. This is a particularly demanding task and the young man still clearly remembers the crouched position and the repetitive, tiring actions required to pull out the roots. In 2011, while studying microtechnology at the Faculty of Business and Economics of the University of Lausanne, Tanner met Aurélien Demaurex, a graduate and talked to him about his idea of using a robot without any chemical weed-killers. Demareux was soon convinced of the merit of such a project. They worked together to develop a concept, at the same time producing a market study which confirmed to them that this was a project with commercial potential. Their company was founded in Yverdon-les-Bains in 2011, with the name ecoRobotix. The first prototype was completed shortly afterwards.This first version of the robot did not use articulated arms to pull out weeds, but instead sprayed a small quantity of weed-killer on an area that had been delimited beforehand. Solar-powered, it was able to work for up to 12 hours without human interaction and to use 20 times less herbicide than traditional methods.

To date, Tanner and Demaurex have designed five versions. In 2018, ten unitshad already been ordered by Grunderco, a Swiss distributor that specialises in selling agricultural equipment. Although ecoRobotix did use weed-killer for the first version of its robot, this was simply because only 10% of Swiss farms are organic.However, the start-up has not forgotten its initial robot project and is due to launch another version in 2019, which will destroy weeds without using weed-killers.

Then there is the *Bonirob* developed since 2014 by Professor Amos Albert, a robotics expert and a team at BoschDeepField Robotics in Ludwigsburg, Germany. Albert had previous worked on Bosch's autonomous lawn mower. The plant appraisal robot, which is approximately the size of a compact car, uses video- and LiDAR-based positioning as well as satellite navigation to find its way around the fields. It knows its position to the nearest centimetre. It also helps minimize the environmental impact of crop farming.*Bonirob*, also known in German as *Bonitur*, can analyse the performance of new plant varieties On the basis of leaf shape, *Bonirob* can distinguish between crops and weeds. With the help of a rod, it gets rid of weeds mechanicallyrather than with weed-killer. Undesired plants are simply and swiftly rammed into the ground.

For further information about autonomous electric agricultural machines, please read Chapter 6.

Appendix B

Electric Lawn Mowers

The use of electric agricultural machinery on the lawns of private estates began in March, 1895 when William John Stephenson-Peach, who taught engineering at Repton School and Cheltenham College, applied for a patent for "an improved combined lawn-roller and portable motor, together with apparatus connected therewith." The patent includes a device for taking up the slack in the electric cable and generally keeping it clear of the machine and the ground. However, there is no mention of lawn-mowing capabilities in this patent.

It was not until 1925 that an American firm called Coldwell began to manufacture a cable take-up reel electric lawn mower after two years of development. It used a patent obtained by Alvin Smith of Newburgh, New York State who applied for a US patent in August, 1925. (This was granted in April 1931 - USP 1,802,358.) At the same time, Otto Charles Walker of Pittsburgh, Kansas obtained a patent for "an electric lawn mower in which the power for operating the blades of the lawn mower is derived from the ordinary house lighting current source from which a suitable conductor is provided which extends to a convenient

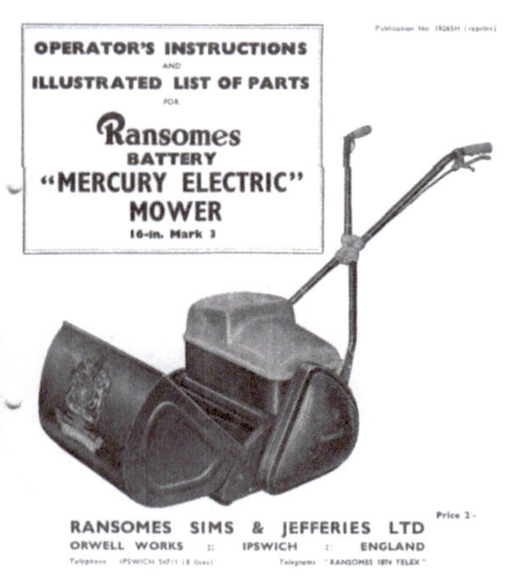

Ransomes, Sims and Jefferies cable lawn mower was launched in 1927. The user was of course limited to the length of the cable. (The Museum of English Rural Life, University of Reading)

point on or near the grass plot to be mowed. This construction obviates the manual toil ordinarily necessary for operating the blades of the mower aside from that work involved in moving the lawn mower across the lawn."

Ransomes, Sims and Jeffries Ltd. of Ipswich, England were already known for their *Orwell* battery electric lorries, refuse tip wagons, works trucks, runabout cranes, prototype electric bus and electric motorcycle with sidecar. In 1926, they produced their *Electra* lawnmower, which was quite widely advertised towards the end of that year, covered by both U.K. and Australian patents. The *Electra*'s most distinctive feature was the methodof using a cable pole to keep the electrical supply cable out of harm's way, though it appears that at least some owners soon removed this tram (street-car) looking feature. The *Electra* was first sold in 41 cm (16 in) and 51 cm (20 in) sizes. The "lighter" and somewhat cheaper 36 cm (14 in) model called the *Countess,*was added to the range a little later. The machines were marketed as *Elektras* in the important German market. The range was available until about 1936. Ransomes then announced further electric lawnmowers , the *Lawnic* and *Bowlic* which had been developed during 1934. The *Bowlic* was advertised as being particularly suitable for golf courses, cricket wickets and bowling greens. The *Electra* was still available after 1945 in various versions, but essentially for the export market.

Ransomes were not alone, the electric *Rotoscythe*, first made by Power Specialities of Maidenhead, later of Slough and then taken over by J. E. Shay of Basingstoke, was the first practical rotary mower. J. E. Shay Ltd., a large concern with major interests in fork-lift trucks, had been founded by Sir Emmanuel Kaye and John Sharp. The inventor of the *Rotoscythe* was David Hamilton Cockburn, a recognised engineer who had already had patents granted to him in other engineering areas before he applied for the first *Rotoscythe* patent on 29th February, 1932. That first patent, GB 385,473, was granted on 29th December, 1932.

Ransomes did produce a battery-operated model and a 240-volt mains machine. The battery model was for domestic use and produced from the late 1950s for about 10-12 years. The machine was known as the *Mercury*. At this time, engine-driven machines were notoriously difficult to start and not particularly reliable, so the attraction to electric propulsion was obvious. The *Mercury* was produced in 36 cm (14 in) and 41 cm (16 in) version with a cheap 12-volt car battery as the power unit. The mower was supplied with a charger and was popular until engine mowers became more reliable in the mid 1960s.

A design by Josephus Miller of Louisville, Kentucky, USApre-dates the *Rotoscythe*. The machine was sold by the Louisville Electric Manufacturing Company under the name *Pioneer*. US patent 1,831,681 was applied for by

Miller on 19th October, 1928 and granted on 10th November, 1931. The machine featured a horizontal cutting blade directly connected to a vertically mounted electric motor. The height of cut of the described lawnmower was adjusted by moving the entire motor/blade assembly up or down as required. The blade was described as being either a single one sharpened at its "diametrically opposed" ends as in many modern rotaries, or as a transverse bar to whose extremities cutting blades or circular cutting disks were attached. The mower seems to have cutgrass adequately, and at least one example apparently still does.

Unlike the *Rotoscythe*, the Louisville mower did not feature any grass collection system nor were the cutting blades completely surrounded by the deck (or otherwise) of the machine. The patent drawings bear a resemblance to the rather later and interesting Australian *Tecnico* electric lawnmower, of which at least one example also exists.

Development has continued.

The automatic lawn mower was invented by S. Lawrence Bellinger and called the *Mowbot.* In 1945, the young Bellinger while working with Norman F. Barnes at the Analysis General Engineering and Consulting Laboratory of General Electric Company, Schenectady, New York had developed a high-speed mercury lamp for flash photography. Scarcely longer than a kitchen match but consuming 4 million watts during its brief flash, it could take pictures of the invisible in four millionths of a second. They were able to take a picture of the air disturbances at the muzzle of a gun at the moment of firing. Known as Schlieren and Shadowgraph Equipment it was used for Air Flow Analysis. In the years which followed Bellinger was working at the Newton Television Centre and then as Head of Spectography at Saratoga Springs Laboratories.

In 1967, S. Lawrence Bellinger invented MowBot, the world's first robotic electric lawn mower, manufactured in Tonawanda, New York (Public Domain).

Then in 1967, Bellinger developed a battery powered, self-propelled random motor lawn mower which used a boundary cable buried beneath the surface to inform the robot where the borders of the lawn were placed. The border wire led to a signal generator which converted ordinary house current to a low voltage 3000 cycle signal to which the control mechanism responded when the mower approached the signal wire. Left and right side sensors were fitted front and rear in conjunction with a solid state electronic controller. Bellinger's electric mower, able to cut up to650 m^2 (7,000 ft^2) on one charge was called the *Mowbot* and was manufactured in Tonawanda, New York. In an attractive yellow livery, the *Mowbot*, priced at $795, was marketed as "the world's only truly automatic lawn mower" with the slogan "It mows while you doze!" In December, 1968 *Popular Mechanics* magazine ran a story "Amazing New *Mowbot*: It really mows your lawn by itself!" In 1971, Bellinger was granted US Patent 870714A. During the 1970s, Bellinger became a pure scientist working for the Technicon Instruments Corporationin Tarrytown, New York where he invented the flow cell for photometric analysis of sample fluids and other instruments for analyzing blood samples, human tissue, and industrial chemicals.

Among those patents which followed, one was taken out in 1993 by a U.S. Republican politician and writer calledNewt Gingrich. It was for a 2WD lawn mower with battery-powered independent electric motors for both driven wheels. The motors are combined with reduction gearboxes, and the wheel is mounted on the output shaft of the gearbox. There is no mechanical axle connection. The power to the motors is controlled by an automatic controller as to wheel speed, power being increased/decreased in accordance with whether the wheels are under-running or overrunning relative to manually-controlled speed setting. Steering is controlled by potentiometer on a mechanical steering wheel or by joystick control. The joystick could be on the tractor or remote.

Over in Israel in the early 1990s, Udi Peless and Shai Abramson, an electronics engineer from Tel Aviv University, had a clear vision: a series of completely autonomous products that would perform our repetitive home chores such as washing the floor and dusting. They founded a company called Friendly Machines in Even Yehuda tomanufacture robotic lawn mowers called *Robomow*. Legend has it that one hot summer's day, Mr. Peless's wife asked him to mow the lawn. While trying to avoid this impending chore, Udi fantasized about a robotic lawnmower. The patent they applied for in 1995 concerned "a navigation method and system for autonomous machines with markers defining the working area". The *Robomow* prototype was displayed at a large gardening exhibition in England in 1997, then first released in Israel in 1998 when mass production and marketing commenced. The brand gained success almost immediately among garden owners thanks to the daily and entirely independent care that it lavished on the home lawn.The product was released in Europe and the U.K. in 1999. That year the company name was changed to Friendly Robotics with its headquarters in Pardesiya, Israel. Since then, the company has continued to innovate and improve their product line. In 2000, the second generation of robotic mowers arrived: the *Robomow RL* platform, more advanced, smaller, lighter and significantly more user-friendly. Further enhancements included the Base Station, enabling users to create a periodic lawn mowing programme. More recently, *Robomow* has developed its own mobile application (the *Robomow* app) for remote and interactive control. *Robomow* is sold in approximately 20 countries and mows about 300 thousand gardens around the world. In 2018, the company was expected to manufacture 50 thousand lawn mowers.

Also in 1995, the Swedish company Husqvarna launched the world's ûrst solar-powered lawn mower, developed by a team led by Belgian electrical engineer André Colens. The product, named *Solar Mower*, was Husqvarna's entry into the robot lawn mower market It was able to find its charging station with the help of emissions from a radio frequency by following a boundary wire.Powered entirely by solar energy, the machine was the predecessor of the Husqvarna *Automower* of which one million had gone into service by 2017.

During the past decade Jason A. Dean Technologies, Irobot, MTD Products, and Honda have also produced innovative units.

Today's solar-powered lawn mower is intelligent enough to recognize you. It comes with added intelligence that lets it identify the route, avoid obstacles and stop it from sliding down a slope. It also comes with image identification and telenet video inspection that lets it work automatically

The manufacturer LG Electronics' Lawn Mower Robot, built in Nashville, Tennessee will work with Amazon Alexa and Google Assistant, the virtual assistants that you can cue with your voice. That means the owner should be able to say, "Alexa, cut my grass" or "OK, Google, mow the lawn" to their Alexa- or Google Assistant-enabled smart speaker when the grass is too long, and the Lawn Mower Robot will automatically do its thing. To do this LG equipped the boxy mower with GPS tracking that collects data about where it needs to go in your yard. The GPS will also let the owner know where their Lawn Mower Robot is if someone steals it from their lawn.But owners must use the Bellinger system to install a perimeter wire around their yard so the robot mower knows where to stop mowing.

As with all other forms of electric transport, the humble cordless lawn mower has also benefitted from the arrival of the lithium battery. A particular innovation is incorporated into the Black and Decker 48cm (1.6 ft) 36V lithium-ion cordless mower, where its Autosense® detects the load on the motor and operates using the optimum power for the task. Performance is increased when more power is needed and the battery runtime is extended during lighter cutting tasks. Other manufacturers include Eco Power, Greenworks and Earthwise.Ransomes-Jacobsen, formerly Ransomes, Sims and Jeffries have been transitioning to zero-emission and diesel-electric mowing equipment bringing the wheel full circle since the *Electra* cable mower was launched in 1927. Ransomes Jacobsen produced the first hybrid-electric walking greens mower in 2007 and the first hybrid-electric riding greens mower in 2009. The *Eclipse® 322* features a hybrid or electric power system, ergonomic design, and the Jacobsen Classic XP™ reels.

Hydrogen fuel cell lawn mower prototypes have appeared. In 2012, industrial designers Otto Polefko and Balazs Botos in Sheffield, England came up with the H²Mow, aka *The Crab*. The compact mower's handle also folds flat on the body for easy storage or transport. The concept produces hydrogen from rainwater whenever it is connected to a wall charger. Hydrogen produced by the system is then stored in three large tanks made using lightweight and strong composite materials that are explosion-proof.In the USA in 2013, Mike Striziki, inventor of the Hydrogen House Project of Hopewell, New Jersey presented the solar hydrogen fuel cell electric ride-on lawn mower. This prototype, an adapted Hustler *Zeon* electric ride on zero-turn lawn mower, built in Hesston, Kansas was upgraded with a renewable energy system capable of refuelling itself, quadrupling the driving range on a single charge and increasing the vehicle's battery life.In France, in 2016MaHyTec,a developer and producer of hydrogen fuel tanks,based in Dole in the Franche-Comté region, created world's first hydrogen-powered *riding* lawnmower. Called the BAHyA, the prototype was sold to a professional gardening company in France. To date a production model has not appeared.

Appendix C

Electric Agriculture Irrigation Machines

For centuries, watering on farms and in orchards was carried out using irrigation channels. The first major irrigation project was created under King Menes during Egypt's First Dynasty between 700-600 BC. He and his successors used dams and canals (one measuring 20 km / 12.4 mi) to use the diverted flood waters of the Nile into a new lake called Lake Moeris. By 1800 AD, irrigated acreage worldwide had reached 8 million hectares (19,760,000 acres). The source of power to drive the pump was an important consideration for the irrigator. Very early irrigators used horse power for shallow wells; later, they adapted steam engines. When, electric power arrived on the farm, irrigators began using large electric motors. As World War II ended, gas, oil and diesel motors were set up to power wells.

Centre-pivot irrigation was invented in 1940 by farmer Frank Zybach, who lived in Strasburg, Colorado. In 1949, Zybach obtained a patent for "A self-propelled sprinkling irrigation apparatus", obtaining US2604389A in 1959:

> Previous driving or rotating mechanisms proposed have included sets of cables wound around drums, for pulling the pipe around various portions of the field, or internal combustion engines *or electric motors at each of the supports*. A rather complicated and cumbersome overhead trolley suspension system extending around the periphery of the field has also been proposed. All of these previous proposals have suffered from numerous disadvantages, such as the need for constant attention or abnormally high initial cost.

In 1954, Zybach licensed his patent to Robert Daugherty and his company, Valley Manufacturing. Daugherty's engineers spent the next decade refining Zybach's innovation, making it sturdier, taller and more reliable and converting it from a hydraulic power system to electric drive.

In 1965, Leo J. Dowd of Columbus, Nebraska received Patent N°3342417Afor a self-propelled irrigation system of the cable type using electric motors. Other centre pivot manufacturers such Lindsay, Reinke, Valley, Zimmatic, Pierce and

Grupo Chamartin used systems driven by an electric motor mounted low on each span. This drove a reduction gearbox and transverse driveshafts transmitted power to another reduction gearbox mounted behind each wheel. In 1968, Richard F. Reinke developed the first variable speed reversible electric drive centre-pivot and in 1977 Valmont produced their lateral or linear-move irrigation machinefed by flexible hoses.

At the same time, the precision innovation of surface drip irrigation was being innovated by Simcha Blass, a Polish-Israeli engineer and inventor. Blass revolutionized drip irrigation in the Yishuv (Jewish community) of Palestine in the early 1930s, pretty haphazardly. A farmer drew his attention to a big tree growing in his backyard "without water". After digging below the apparently dry surface, Simcha Blass discovered why: water from a leaking coupling was causing a small wet area on the surface, while an expanding onion-shaped area of underground water was reaching the roots of this particular tree, and not the others. This sight of tiny drops penetrating the soil causing the growth of a giant tree provided the catalyst for Blass' invention. Blass made tubing that would release water slowly and steadily through larger and longer passageways, using friction to keep the flow. In the late 1950s, with the advent of modern plastics during and after World War II, Blass took a major step towards implementing his idea. With his son Yeshayahu, they developed a plastic emitter.The first experimental system of this type was established in 1959 by Blass who partnered later (1964) with Kibbutz Hatzerim to create an irrigation company called Netafim (Hebrew for "droplets."). Today, Netafim operates in over 100 markets worldwide through 29 subsidiaries,employing over 4,500 people in 17 manufacturing plants.

The next step was to link solenoid-valve micro-drop irrigation to Landsat imagery, available from the 1970s, the best and least expensive way to quantify and locate where water is used and in what quantity (see Appendix D). In 2003, Landsat ET estimates helped to increase the flow of the Yakima River, Washington State while maintaining the monetary level of crop production. In New Mexico, Landsat ET maps helped water managers strike a balance between irrigation demands and riparian vegetation requirements. And in California, Landsat helped create a state-wide water use plan to enable farmers to determine their precise irrigation needs.

The latest step is Artificially Intelligent Micro Irrigation (AIMI). In 2010, S. Muhammad Umair and Rasyidi Usman from National University of Science and Technology in Pakistan proposed to create an automated irrigation system using an Artificial Neural Network based controller. The input parameters such as air temperature, soil moisture, radiations and humidity are modelled. Then using an appropriate method taking into account ecological conditions, evapotranspiration and type of crop, the amount of water needed for irrigation is estimated and then associated results are simulated.

To improve water efficiency for blueberry cultivation, in 2017, researchers from Chile's Universidad Católica de la Santísima Concepción (UCSC) created an AIMI (Artificially Intelligent Micro Irrigation) to increase water efficiency by 70%. The National Centre for Engineering in Agriculture, University of South Queensland of Australia has developed software tools and hardware technologies to improve the measurement, evaluation, optimisation and control of key inputs for both manually operated and automated irrigation and fertiliser application systems in cotton cultivation.Micro-drip pioneers Netafim in Israel have launched *NetBeat,* a system providing farmers with real-time recommendations sent to a mobile app, based on data gathered from sensors, including data about plants, soil, and weather, as well as from satellite images and weather forecasts. Aadith Moorthy at ConserWater has innovated the ConserWater Web, a web dashboard, in the US so no matter whether they grow almonds in California or peaches in Georgia, farmers now use a simple dashboard interface to water crops efficiently.

While AIMI has become an integral and vital part of today's precision farming, drone irrigation is an efficient ally for inaccessible zones. *(See Chapter 5)*

Appendix D

Satellite Farming

Man's fascination with flight launched many dramatic adventures in the early days, some ending in disaster as gravity won over ingenuity. "La FuméeElectrique" (= "Electric Smoke") was the phrase erroneously used by the Montgolfier brothers, Joseph-Michel and Jacques-Etienne, to describe the hot smoky air whichin the summer of 1782, raised their pioneering balloons to a height of 300 m (980 ft) into the skies above Annonay, in France's Ardeche region. The brothers had begun experimenting with balloons made of paper and early experiments using steam as the lifting gas were short-lived due to an effect on the paper as it condensed. Mistaking smoke for a kind of steam, they began filling their balloons with hot smoky air which they called "electric smoke". The French Academy of Sciences soon invited them to Paris to give a demonstration. On 19thSeptember, 1783 the Montgolfier brothers' balloon recorded the first successful hot-air balloon flight over Paris. A sheep, a duck and a rooster become the first passengers. The balloon rose to about 1,829 m (6,000 ft) and landed safely.

On 12thDecember, 1783 Horace-Bénédict de Saussure published a letter in the *Journal de Paris* to prove that:

> …contrary to what scientists are thinking, it is not the "lightness" of this air pushing up aerostatic machines; but the air rarified by the heat of the flames. The evidence is very easy to check: by introducing a piece of metal white hot in a paper bag, it is easy to propel it to the ceiling. The strong smell of electrical smoke remains!

Prior to the development of heavier than air flying machines, gentlemen-farmers used Montgolfier balloons to inspect the state of their crops and their forests.

In that satellites,orbiting at altitudes above 2,000 km (100–1,240 miles), usesolar panels to energise their electric battery, so enabling them to carry out crop monitoring, this defines them as EAMsIt is essential for conservationists, governments and landowners to inventory forests for ecological, social, and economic health.

Consider a farmer riding along in their20,000 hectare (50,000-acre) wheat field early in the growing season. He pushes a button on his tractor to turn on its satellites-based Global Positioning System (GPS) monitor, which pinpoints his exact location to within one metre. Touching another button, he displays a series of Geographical Information System (GIS) maps that show where the soil in the field is moist, where the soil eroded over the winter and where there are factors within the soil that limit crop growth. Next, he uploads remote sensing data, collected just yesterday, that show where his budding new crop is already thriving and areas where it isnot. He hits SEND to upload these data into an onboard machine that automatically regulates the application of fertilizer and pesticides, just the right amount and exactly where the chemicals are needed. He sits back and enjoys the ride, saving money as the machines do most of the work. This is a precision farmer.Various developments over the years in satellite positioning, observation and the use of energy have been put together to allow for this new type of technological agriculture.

In 1957,the Soviet Union launched *Sputnik*, the first artificial satellite to orbit around the Earth. It had a mass of 83.6 kg (184 lb) and travels in an elliptical orbit at a height above the Earth between 215 km (134 mi) and 939 km (583 mi). It travelled at 29,000 km/h (18,000 mph) and takes 96.2 minutes for each orbit.Two American physicists, William Guier and George Weiffenbach at Johns Hopkins University's Applied Physics Laboratory (APL) decided to monitor its radio transmissions. Within hours they realized that because of the Doppler effect, they could pinpoint where the satellite was along its orbit.

The U.S. Department of Defense was swift to realise the important part solar panels could play as a back-up power source. When the initial 1.4 kg (3 lb) spherical *Vanguard* satellites were launched in the late 1950s, they contained as their payload seven mercury cell batteries in a hermetically sealed container, two tracking radio transmitters, a temperature sensitive crystal and six clusters of solar cells on the surface of the sphere. The latter allowed *Vanguard 1* to continue transmitting for over a year after its chemical battery was exhausted. The successful operation of solar cells on this mission was duplicated in many other Soviet and American satellites, for example in 1962, *Telstar*, the first communications satellite, was powered by 3,600 solar batteries.

At the same time another technology was developed that would go arm in arm with satellite observation. LiDAR originated in the early 1960s, shortly after the invention of the laser, and combined laser-focused imaging with the ability to calculate distances by measuring the time for a signal to return using appropriate sensors and data acquisition electronics. Its first applications came in meteorology, where the National Center for Atmospheric Research used it to measure clouds. The general public became aware of the accuracy and usefulness of LiDAR systems in 1971 during the *Apollo 15* mission, when astronauts used a laser altimeter to map the surface of the moon. The interpretation of "lidar" as an acronym (LIDAR or LiDAR) came later, beginning in 1970, based on the assumption that since the base term "radar" originally started as an acronym for "Radio Detection And Ranging", "LIDAR" must stand for "Light Detection And Ranging". A LiDAR instrument principally consists of a laser, a scanner and a specialized GPS receiver.

Valerie Thomas Landsat image processing in 1972. Thomas was one of the image processing specialists who facilitated the ambitious Large Area Crop Inventory Experiment, known as LACIE—a project that showed for the first time that global crop monitoring could be done with Landsat satellite imagery. (NASA)

In 1978 launched from an Atlas E/F SGS-1 rocket, a succession of OPS 511 satellites pioneered the GPS system which has since become so crucial to precision farming.

On 23rd July, 1972, the Earth Resources Technology Satellite was launched. This was eventually renamed *Landsat*. One of those working at NASA's Goddard Space Flight Center, was Valerie L. Thomas, an African-American scientist and inventor. Ms. Thomas' task was to manage the development of early *Landsat* image processing software systems and became the resident expert on the Computer Compatible Tapes, or CCTs, that were used to store early *Landsat*imagery. In doing so, Thomas was one of the image processing specialists who facilitated the ambitious Large Area Crop Inventory Experiment, known as LACIE, a project that showed for the first time that global crop monitoring could be done with *Landsat* satellite imagery. In 1974, Thomas headed a team of approximately 50 people for the Large Area Crop Inventory Experiment (LACIE), a joint effort with NASA's Johnson Space Center, the National Oceanic and Atmospheric Administration (NOAA) and the U.S. Department of Agriculture. LACIE demonstrated the feasibility of using space technology to automate the process of predicting wheat yield on a worldwide basis. Thomas was thus responsible for managing the development of an image processing system and the operations process, with a production rate of 100 test sites per day; coordinating with a contact at another NASA centre and presenting GSFC's LACIE results at a multi-Agency group (NASA/GSFC, NASA/MSFC, NASA/JSC, NOAA and US Dept. of Agriculture). NASA's *Landsat* programme currently has two satellites in orbit imaging the whole of Earth's surface every 16 days. Such information was to prove invaluable to farmers and agronomists worldwide.

GPS is a space-based satellite navigation system that provides location and time information in all weather, anywhere on or near the Earth.The GPS project was launched by the U.S. Department of Defense in 1973 for use by the United States military and became fully operational in 1995. It was allowed for civilian use in the 1980s. It was only with the deployment of Global Positioning Systems (GPS) in the 1980s, allowing the precise positioning of aircraft that made airborne LiDAR surveying possible. Bythe early 1990s, GPS was accurate to within a few feet, good enough to create crop yield maps, which helped farmers make decisions about drainage, weed control, fertilizer and seeding. However, the technology was costly. Precision agriculture, which uses signals from the array of GPS satellites for farming applications, showed promise. But at the time, there were far more questions than answers.

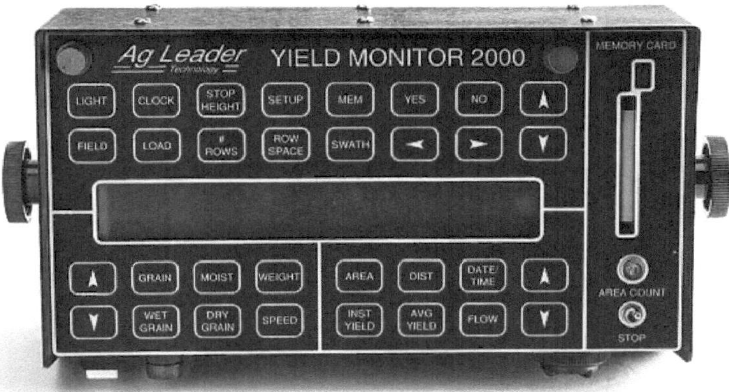

From 1993, by linking the yield data provided by Al Myers' Crop Seed Monitor to GPS-plotted locations, farmers began creating detailed yield maps. (Ag Leader)

In 1992, Allen Myers of Ames, Iowa perfected an on-the-go crop yield monitor, a major step in making precision farming possible. Spot measuring the harvest gave farmers new information about the strongest and weakest portions of their fields. By linking the yield monitor data to GPS-plotted locations, farmers began creating detailed yield maps.Myers worked for six years to develop the crop monitor, working nights and weekends outside of his work at his full-time job. He taught himself electronics and microprocessor programming, and conducted secret field trials on his father's farm in Illinois. Myersthen lefthis job as a design engineer to start a company called Ag Leader Technology. Although in his first year he sold only 10 yield monitors, by 1995 he had sold 1,500 units. Dennis Friestacquired Yield Monitor 2000 N° 15, for his farm in Radcliffe, Iowa. "I always had a desire to keep track of my yields. When I started test plotting different varieties, and weighing them across the field, this technology was something I knew would help me a lot."

John Deere also saw the potential in these systems. In 1994, the firm created the Precision Farming Group to look for evolving technologies that could make GPS more accurate and cost effective for providing location information to create yield maps. At that time, no one had proven that GPS could be used for vehicle guidance. Soon after, a research effort by the John Deere Product Engineering Center (PEC) and the Precision Farming group initiated a project with Stanford University, which targeted an end goal of developing an autonomous tractor controlled by a sophisticated GPS system. John Deere provided Stanford with a tractor modified with electronic valves to control the steering, brakes and transmission. Stanford outfitted the tractor with the GPS system, an on-board computer, sensors to detect obstacles and a telemetry system to communicate with a remote base station.

The early crop yield monitoring systems were rather cumbersome compared to today's compact tablet computer.

In 1994, engineer Terry Pickett argued that Deere should further invest in GPS-based precision agriculture research. Company executives agreed, hoping to encourage farmers to buy new, more-efficient equipment. In 1996, John Deere launched its GreenStar Precision Farming System. Their brochure predicted that "Information is your new crop!"John Deere's first production GPS receiver, nicknamed "green eggs and ham," brought satellite control to the tractor cab. Farmers eventually began to use GPS to steer their equipment, avoid spraying the same spot twice and to discover exactly which areas of a field produced the most.

By this time, the first GPS array consisting of 24 geosynchronous satellites became operational for use by the military. In 1996, recognizing the importance of GPS to civilian users as well as military users, U.S. President Bill Clinton issued a policy directivedeclaring GPS a dual-use system and establishing an Interagency GPS Executive Board to manage it as a national asset.

Stanford University, continuing to work with Bob Mayfield, an engineer with the John Deere Advanced Tractor Engineering group in Waterloo, Iowa, looked at the possibility of automatically guiding a tractor in the field. The technology to take GPS accuracy down to the 2-to4-centimetre(0.8-1.6 in) level existed, but it took two PhDs to support it due to its sheer complexity.It became clear that the linchpin of a successful guidance system would be a simple GPS system supported by the John Deere dealer network. So, while Mayfield continued to work with Stanford engineers to refine the overall product concept, Nelson, Pickett and their advanced engineering peers went back to the drawing board. Teamed with talented engineers from NavCom Technology, now a wholly owned subsidiary of John Deere, and NASA's Jet Propulsion Laboratory, the group spent weeks deconstructing problems and looking at new ways to put the pieces back together. One of these challenges was differential correction, the ability to fine-tune GPS signals by using a ground receiver to adjust for error. Without accurate correction, readings could be off by half the length of a football field.[1] In early 1997, during a demonstration for

In 2000, these were the new tools available to farmers for a combine-mounted yield-monitoring system. *Clockwise from left*, Motion sensor, yield monitor (nicknamed "green eggs and ham), global positioning system, yield-monitor display and data-recording device. (Composite courtesy of John Deere Company).

John Deere management, the specially-equipped tractor made perfectly straight beds with an accuracy of 2.5 cm (one inch), raised and lowered an implement and turned, all with no operator on-board. Another test proved the tractor's ability to till a field without damaging buried irrigation tape. After this demonstration, the team knew the technology would significantly reduce costs and labour requirements.

At the same time, the Precision Farming group teamed with the PEC's Advanced Tractor Engineering Group to create a plan for automating steering, for not only tractors but all other platforms. While it was possible to programme a tractor to perform tasks the tractor still needed an operator.A late-1997 prototype demonstrated how a tractor could automatically steer on a straight line entered by an operator. Known today as AutoTrac, this guidance system is used on many of John Deere's self-propelled products. A key finding of this early work was a need to develop a new GPS receiver capable of supporting the accuracy that a vehicle guidance system needed to satisfy a global customer base from both a cost and performance basis. The John Deere engineering team solved its technical riddles and the results proved dramatic. In 1998, the company first offered satellite-based DGPS systems with one to two metres accuracy (3.3 to 6.6 ft). By 2004, accuracy had increased to 10 centimetres (4 in) and today the numbers continue to shrink.

In 1999, the Stanford-developed system was refined to be more compact and user-friendly. To calm concerns about the system's ability to plant and later cultivate the same ground, an Iowa co-operator used the system to plant 101 hectares (250 acres) of corn and drill 101 hectaresof soybeans. Weeks later, when he drove the tractor into his fields to cultivate the standing crops, he engaged the automated steering system. The tractor precisely followed the same rows that were planted weeks earlier.John Deere AutoTrac™ went into production in 2002. The guidance system has become so popular that today 60-70% of the crop acreage in North America is farmed using AutoTrac or similar systems. In Australia, that number exceeds 90%.

The Smithsonian Institution in Washington, D.C. recently showed interest in learning more about the original GPS receivers. Peter Liebhold, Curator of the Smithsonian Institution's Division of Work and Industry, called the original Deere GPS receiver the "steel plough of the 20thcentury."[2]

Other tractor manufacturers have followed suit, for example New Holland with their *IntelliSteer* guidance system. In Europe since 1996, Avalis' Plant Institute, Terres Inovia and Airbus Defence and Space have pooled their expertise and tools to develop FARMSTAR.In Germany, from 1999, the DEMMIN (Durable Environmental Multidisciplinary Monitoring Information Network) soil measurement and validation site was set up east of the small town of Demmin in Mecklenburg-West Pomerania. Since 2011, it has been equipped with further measuring instruments by the GFZ.Covering more than 30,000 hectares (74,000 acres), the site entertains an environmental monitoring network that is unique in Germany. Nearreal-time information about climate and soil are collected via a radio network in high time resolution. The environmental sensing network comprises 43 stations. It measures among others; air temperature, precipitation, wind and radiation parameters in 15-minute intervals. Additionally, 63 soil moisture gauging stations are installed below the agricultural fields.

Another pioneer of satellite-assisted farming is Susan Moran. Having obtained her PhD in Soil and Water Science from University of Arizona in 1990, Moran became a research hydrologist with the U.S. Department of Agriculture, Agricultural Research Service U. S. Water Conservation Laboratory and also a member of the NASA *Landsat 7* Science Teamand EO-1 Validation Team, based in Tucson, Arizona. Moran developed a method to help farmers with resource management by combining *Landsat* images with radar data from several polar orbiting satellites. Although *Landsat saw* the surface very clearly, its usefulness was limited because it could not see through clouds, and it only flew over a particular area once every 16 days. Most farm management decisions required information on a daily or biweekly basis. Radar can see through clouds and be collected daily, but the resolution is not as good as *Landsat*. Moran worked out a way to combine radar and *Landsat* information to provide a continuous record of the land surface and vegetation health. Her methods were successfully put to the test by land managers through a Cooperative Research and Development Agreement with a commercial image supplier. Moran also pioneered remote sensing of soil and vegetation in cropland and rangeland, providing a guide for regional water management which has been used for irrigation scheduling in Spain, forest fire risk assessment in France, drought monitoring in Africa, and estimating grassland degradation in China. Her leadership in inter-agency collaborations range from work on mission science teams with the National Aeronautics and Space Administration to a report on impacts of extreme weather patterns on grasslands and forests with the U.S. Forest Service. In 2002, Moran, with87 research works, 4,015 citations and 12,992 reads, was honoured as Outstanding Senior Research Scientist of the Year by USDA ARS.

Meanwhile, research and development of LiDAR applications in agriculture have further assisted with mapping water flow and catchments and monitoring erosion and soil loss.As used within forestry, 3D vegetation models can also be created to assist with crop planning and assess patterns such as canopy height, even in dense woodland.LiDAR technology can also be used to collect data that can identify the exact soil type that a certain farmland has. This information is important to the farmer because it helps the farmer know which type of crops can be grown on that farm and what fertilizer should be applied. LiDAR data is instrumental in precision agriculture. LiDAR can also be used to carry out general crop analysis and determine the suitability of the crop to thrive in a particular area. This can be done through estimation of the crop quality and measuring this against the ideal standards.

In 2009, NASA Earth Observatory teamed up with the U.S. Department of Agriculture to use its *Aqua* satellite to monitor soil moisture around the globe. Crop yields are determined in part by how much water the ground holds. The satellite collects data on the amount of microwave radiation emitted by the land, which scientists combine with information about vegetation cover and soil temperatures to figure out how much of the radiation is coming from water in the soil. The USDA's Foreign Agriculture Service is using the data to help form more-accurate crop forecasts for the U.S. and for developing countries, where ground data is often sparse. In 2013, the project received a boost from the Soil Moisture Active and Passive satellite, which has been providing information at a much more detailed scale than *Aqua*.

In 2016, Astro Digital launched the first of its *Landmapper* satellites on a SpaceX rocket from Vandenberg Air Force Base near Lompoc in California. The company's set-up would eventually consist of 30 satellites orbiting 650 kilometres (404 mi) above Earth.A third of these would take images every day at a resolution of one pixel for every 22 square metres (237 ft^2) of land. The rest would capture images at a resolution nine times higher every three to four days. The company's main focus will be on scanning agricultural land, however. South African start-up, Farm is ready to help farmers make the most of Astro Digital's tools, having decided that using drones as a means of monitoring land and crops involving training pilots and maintaining aircraft would be impractical. In China, the Changguang Satellite Technology Co. Ltd. is using the*Jilin-1* satellite for agriculture and forestry in northeast China's Jilin Province.

Last but not least there are nanosatellites. The idea of producing smaller satellites dates back to 1999, when Jordi Puig-Suari at California Polytechnic State University (Cal Poly) and Bob Twiggs at Stanford University asked their students to learn about miniature design and spacecraft design. The students developed specifications to promote and develop the skills necessary for the design, manufacture, and testing of small satellites intended for low Earth orbit (LEO) that perform a number of scientific research functions and explore new space technologies. They called these CubeSats. In the 2010s, advances in computing and the miniaturization of components, driven by the democratization of the smartphone, make this idea technically viable.

A CubeSat is made up of multiples of 10 cm x 10 cm x 11.35 cm (4 in x 4 in x 4.5 in) cubic units. CubeSats have a mass of no more than 1.33 kilograms (2.9 lb) per unit, and often use commercial off-the-shelf (COTS) components for their electronics and structure. CubeSats are commonly put in orbit by deployers on the International Space Station, or launched as secondary payloads on a launch vehicle.

2014: William Marshall and Robbie Schingler and a close-up view of a Flock 1 nanosatellite (image credit: Planet Labs)

In 2011, Puig-Suari and Scott MacGillivray, former manager of nanosatellite programmes for Boeing Phantom Works, established Tyvak Nano-Satellite Systems in San Luis Obispo, California to sell miniature avionics packages for small satellites, with the goal of increasing the volume available for payloads. At the same time, Chris Boshuizen,Will Marshall and Robbie Shingler, former NASA engineers,began building a modified Triple-CubeSat in a garage in San Francisco. In April, 2013Planet Labs as they called their start-up, launched two demonstration CubeSats, *Dove 1* and *Dove 2*, aboard a *Soyuz* rocket placing them in a sun-synchronous orbit. *Dove 3* and *Dove 4* were launched seven months later, by which time Planet had announced plans for *Flock-1*, a constellation of 28 Earth-observing *Doves*. Each *Dove* is equipped with a high-powered telescope and camera programmed to capture different swathes of Earth. Two hundred cameras thus orbit around the globe and take some 1.3 million high-definition pictures every day, covering the entire earth's surface. The *Flocks*are designed to obtain 3 to 5 m (10-16 ft) high resolution images of Earth, collecting images from latitudes that are within 52 degrees of Earth's equator. A large portion of the world's agricultural regions and population lie within that area. Planet currently operates the largest fleet of satellites in the world. As of January 2018, about 2,000 Cubesats had been launched by schools, companies, government agencies and organizations worldwide, over 900 of them successfully deployed in orbit, though over 80 have been destroyed in launch failures. In one week, 50 were launched.

In 2018, Planet partnered with Granular, a software company that provides farmers with tools to reduce risk and increase returns, Granular seeing the benefits from high-definition images taken from the sky, which it can integrate into its analysis tools to obtain more accurate predictions. Granular was founded in 2014, and backed by leading venture capital investors such as Andreessen Horowitz and Google Ventures, designed the leading farm management solution, now called Granular Business, by collaborating with many of the largest, most progressive family farms in the country. In 2017, Granular joined forces with DuPont Pioneer, coupling the strength of Granular's business software with DuPont's robust agronomy expertise and the Encirca agronomy software and advisory services. In February, 2018 DowDuPont™ announced a spin-off of the Agriculture Division, naming the new agriculture company Corteva Agriscience™ (which is derived from a combination of words meaning "heart" and "nature.") Granular is an independent subsidiary of Corteva and is responsible for all software and analytics products.

The global micro-satellite market, as manufactured by 26 different aerospace companies, is expected to grow from USD 2,635.64 million 2017 to USD 8,463.65 million by the end of 2024.

Bigger sister satellites still have their placein playing a vital part in precision farming around the world. NASA's ECOSTRESS (The ecosystem Spaceborne Thermal Radiometer Experiment on Space Station) is designed to gauge how much water plants need when they are growing under favourable conditions. ECOSTRESS, nicknamed the "Space Botanist" can address three overarching science questions: How is the terrestrial biosphere responding to changes in water availability? How do changes in diurnal vegetation water stress impact the global carbon cycle? Can agricultural vulnerability be reduced through advanced monitoring of agricultural water consumptive use and improved drought estimation? The ECOSTRESS radiometer was built at the Jet Propulsion Laboratory, and delivered to the ISS by the Space X *Dragon*. The *Dragon* arrived at the space station on 3rdJuly, 2018. The radiometer was mounted on the station's Kibo module. It soon began collecting data and by August it was sending back imagery of three wildfires burning in California and Nevada, following up with variations in surface-temperature patterns in Los Angeles County.

Endnotes

[1]*Jill Brimeyer,* "Nothing Runs Like a Precision Farming System", Progressive Engineer, March 2005.

[2]Brian Holst, "John Deere's GPS: "The steel Plough of the 20th Century" The Plowshare, History for John Deere Collectors Issue #40.

Appendix E

RFID

Radio Frequency Identification technology identifies an object by radio frequency without any contact.It uses electromagnetic fields to automatically identify and track tags attached to objects. The tags contain electronically stored information. Passive tags collect energy from a nearby RFID reader's interrogating radio waves. Active tags have a local power source (such as a battery) and may operate hundreds of metres from the RFID reader. From the early days, RFID was used to monitor cattle and more recently plant traceability.

In 1969, Mario W Cardullo an American inventor, applied for a patent for a passive, read-write radio-frequency identification tag, US3713148A, which was granted on 23rdJanuary, 1973.Cardullo recalls:

> In 1969, I was the corporate planning officer to chairman of the Communications Satellite Corporation (Comsat). In the spring of 1969, I was seated next to an IBM engineer on a flight to Washington, from St. Paul. The engineer was implementing the CARTRAK optical system for the railroad industry. This system consisted of a reflective colour bar code placed on the side of each railroad car. As the railroad car passes an optical base station, the station would transmit a beam of light. The optical bar code would reflect back a signal associated with the individual car so it could be identified.After the IBM engineer finished talking, I started to sketch in my notebook the idea for the RFID tag with a changeable memory. The original sketch showed a device with a transmitter, receiver, internal memory, and a power source. I signed and dated the sketch and went back to reading a book I had started.[1]

The initial device was passive, powered by the interrogating signal, and was demonstrated in 1971 to the New York Port Authority and other potential users. It consisted of a transponder with 16 bit memory for use as a toll device.

The U.S. government was also working on RFID systems. In the 1970s, Los Alamos National Laboratory was asked by the Energy Department to develop a system for tracking nuclear materials. A group of scientists led by Dr Jeremy Landt, came up with the concept of putting a transponder in a truck and readers at the gates of secure facilities. The gate antenna would wake up the transponder in the truck, which would respond with an ID and potentially other data, such as the driver's ID. This system was commercialized in the mid-1980s when the Los Alamos scientists who worked on the project left to form a company to develop automated toll payment systems. These systems have become widely used on roads, bridges and tunnels around the world.

At the request of the Agricultural Department, Los Alamos also developed a passive RFID tag to track cows. The problem was that cows were being given hormones and medicines when they were ill. But it was hard to make sure each cow got the right dosage and was not given two doses accidentally. Los Alamos came up with a passive RFID system that used UHF radio waves. The device drew energy from the reader and simply reflected back a modulated signal to the reader using a technique known as backscatter. The 1970s were characterized primarily by developmental work, including intended applications for animal tracking. Interest in animal tagging was also high in Europe where Alfa Laval, Nedap and others were developing RFID systems.

After developing the technology for government use in the 1970s, the team of Los Alamos National Laboratory scientists moved this technology to the private sector in 1983 when they founded Animal Management Technology (Amtech) Corporation. In addition to Santa Cruz-based Identronix Research, to explore RFID commercialization opportunities. Amtech, targeted the cattle industry. Kevin Ashton led pioneering work on RFID networks, for which he coined the term "The Internet of Things."

RFID technology is part of the arsenal of electric agricultural machinery in the use of its biocompatible glass tags, equipped with an OTP microchip. Thereby, several data rates can be programmed and data can be encoded via configuration Word in Electrically Erasable Programmable Read-Only Memory (EEPROM). Read and write access to EEPROM can be protected with a 32-bit password and a unique ID. In this way, farmers can use it in areas such as environmental monitoring, irrigation, specialty crops, livestock and farm machinery. One example of this involves RFIDs placed in woody plants to store and retrieve information on their health status through all phases of propagation and in the field. The microchip is linked to a database in which many other kinds of information such as pesticide applications, can be collected and linked. Using a Web-based platform, information can be shared globally and accessed quickly. RFID technology can also be integrated with cell phones and netbooks for the easy recording of images and audio, which can be linked back to the chip and shared, or with global positioning systems (GPS) used to create a virtual orchard or vineyard. There are myriad uses for this new technology, which is expanding rapidly and has been implemented successfully in the European livestock industry. The use of RFID technology in agriculture aims at guaranteeing accurate traceability from the farm to the final consumer.

One example of RFID technology took place in Greece. Yiannis Ampatzidis earned his PhD in agricultural engineering from the Aristotle University of Thessaloniki. He received his Master's and Batchelor's degrees from the same university. In 2005, he won a competition in the specialty "Automations in Precision Agriculture". From 2009, Ampatzidis was a researcher in National Agricultural Research Foundation (NAGREF) Institute of Soil Science in Thessaloniki. That year, with several colleagues, Ampatzidis showed how RFID technology could be used to overcome the limitations of existing yield mapping systems during manual fresh peach and kiwi harvesting in two different fields at the village of Episkopi, near the town of Veria, in the region of Macedonia in Northern Greece.

They used two methods to automatically match bins containing harvested fruits with corresponding trees during harvesting in orchards where GPS data may be unavailable due to foliage. Both methods usedlong-range radio frequency identification (RFID) antenna located on the harvesting platform for tree identification. Bin registration was accomplished in the first method by passive RFID tags attached to the bins, whereas the second method used a barcode reader located on the platform, and low-cost barcode tags on the bins. Additionally, a digital scale was used with both methods to measure the yield distribution in the field during the loading of the bins. During the experiment, the aim was to estimate the tree and bin detection accuracies of both methods and their effect on the bin loading time. Statistical analysis of the data showed that when compared to the current standard harvesting procedure, RFID bin registration did not affect the amount of time to stack a bin on the platform (loading time), whereas barcode bin registration increased this time by 14%.

This is just one example of RFID in EAM technology. Many other applications have been used since.

Endnotes

[1]Mario Cardullo, Genesis of the Versatile RFID Tag, *RFID Journal*.

Appendix F

Electric Aquatic Agriculture Machines

Farming of fish in ponds is an ancient practice. It was presumably developed by early farmers as one of many primary production systems to stabilize food supplies. Ancient manuscripts and hieroglyphs have shown evidence of freshwater aquaculture from approximately 3,500 years ago, both in China and in Mesopotamia. Around 500 BC, the Romans farmed oysters and fish in Mediterranean lagoons. The integration of pond fish culture and rice farming is documented as early as the Mid-Eastern Han Dynasty (AD 25-220) in China. Farming domesticated carp in the ponds of Christian monasteries in Central Europe became part of the food production system as did mussel farming from the 13th century. The aristocracy were the main users of freshwater fish *vivaria*, since they had a monopoly over the land, forests and water courses.

Aquaculture today is the farming of aquatic organisms in both coastal and inland areas involving interventions in the rearing process to enhance production. Harvesting is a complex operation involving driving and concentrating the fish stock to the catch basin;electric grids or electrified seines can also be used for driving, where the electric field forces the fish to swim towards the catch basin. In order to achieve more accurate grading, there are graders where the fish does not merely slide down on the slope of the table by gravity but is led by an electric motor-driven endless rubber belt with soft rubber fingers above the slot.

Electrically-powered paddle wheel aerators are a fairly recent innovation. In 1969, Edward J. Baumann of the Beloit Passavant Group obtained a patent for a floating aeration motor "for use in a basin and having a pair of bladed rotors rotatably mounted in a frame. The blades are powered so as to propel the front blade about the basin and the rotors are guided over a predetermined path on the water. Pier 32 and cable 34, in addition to guiding the rotor also serve to support the electrical connections which connect motor 24 with a source of power on the lagoon's shore."

These cable-operated high-quality 7.5 kW paddlewheels usually feature high-efficiency 60 HZ motors and high-quality gear reducers to ensure a long service life. The frame, drive shaft and hardware are in stainless steel, all compatible with salt water. Alibaba.com currently offers 542 solar paddle wheel aerator products. About 90% of these are aquaculture machine aerators. Some Paddle Wheel Aerators are solar-powered. The eO2 capacity of 420 watts of high efficiency, lightweight solar panels, coupled to a 24v DC motor enables this aerator to be sited anywhere on a lake providing instant aeration during daylight hours.

With the advent of electricity, one or two experiments were tried out for fishing. In 1888, in Pittsburgh, Pennsylvania, Louis Semple Clarke or "LS", the 22-year-old innovative photographer and member of the three-year-old South Fork Dam Fishing and Hunting Club on Conemaugh Lake, had come up with an ingenious idea for fishing. Onto a twin-hulled catamaran he mounted a homebuilt electric motor and a battery to drive a 25 cm (10in) screw at 460 rpm. Admired by club members and friends, Andrew Carnegie, Andrew Mellon and Henry Clay Frick, wearing a sailor's suit, "LS" silently cruised out into the middle of the lake where he would cast his line. He re-charged his electric catamaran's battery from the 8-light dynamo on his father's steam launch. He even fitted it with a searchlight on the front. As the craft was powered by electricity, it quickly attained the nickname *Sparky* or *Old Sparky*.

Three years later in Paris, Gustave Trouvé, battery-electric vehicle and apparatus pioneer patented a deceptively simple device for automatic fishing. He conceived of a net around the edge of which was inflatable India-rubber tubing which could be used to lower and raise the net from a distance. The inflation and deflation of the tubing was carried out by means of a switch-controlled compressed air generator which was either on board the fishing boat or on shore. The inflatable net was also attached to a buoy which carried an electric lamp fitted with a reflector so the fisherman could see his net at night. In addition, the tinkling of a bell attached to this device indicated the entry of fish into the net.Then by using a balance linked to the net, the fisherman could gauge the amount of fish he had caught that night. Several years later Georges Dary would write: "The idea of using electricity for fishing and of attracting fish by submerging electric light dates back several years. But it's impossible to give a definitive decision about the excellence of this process, sometimes successful, sometimes useless."[1]

Offshore, as explained in Chapter xx of this book the lithium-battery energy density has inspired several fishing vessel manufacturers to go electric. In 2016, the 56 metre (184ft) freezer/longliner *Northern Leader*, working from out of Kodiak, Alaska trialled a joystick controlled, eco-friendly propulsion system that runs on electricity. In Norway, *Karoline*, aka the *Selfa Artic Elmax 1099*, has a Siemens 195kW battery pack on board as well as an electric propulsion motor. Whilst being docked, the vessel is charged via a 63 amps 220 V course, with a normal charging time 6-8 hours. Following testing in Tjeldsundet, Norway in September, 2015 *Karoline*was placed in Tromso, Norway to be included in the daily operations of Øra AS.

Another e-fishing vessel has been operating from Urk Harbour in the Flevoland Province of the Netherlands: *MDV1 Immanuel* is a revolutionary new 30metre (100 ft) fishing trawler. Built by Hoekman Shipbuilding for the Romkes and Hendrik Kramer families, she is part of the Dutch fishing industry's "Duurzame Visserij" sustainability fisheries masterplan. Several years ago in the Dutch flatfish fishery in the North Sea, catching one kilogram (2.2 lb) of fish used up nearly four litres (almost a gallon) of gas oil. *MDV1 Immanuel* will consume less than half a litre (0.9 pint). Her 600 kW propulsion consists of one 500 kW Oswald permanent magnet electric motor turning at 140 rpm and one 100 kW diesel genset. At the end of a perfectly horizontal propeller shaft, her three-bladed wing-profile Kort nozzle rudder-propeller drives measure 3 m (10 ft) in diameter. The "schroefrende" aqua-dynamic bow and hull shape, and a coating of Sigma Glide anti-fouling further reduce resistance. Even the novel fishing technique where the nets float just above the bottom requires much less boat speed and creates less disturbance to the seabed.

There is also research into the twin rig pulse technique, an unprecedented fishing technique consisting of a trawl net with mild electrical impulses when the fish respond. After three years of reliable service, Padmos in Stellendam and Urk-based Hoekman Shipbuilding have reportedly received thirteen orders since the first vessel came into service in early 2016. At the time of writing, Padmos has built six vessels in the MDV range, with seven more currently in production, according the Stellendam shipbuilder.[2]

*MDV1 Immanuel*designer, Frans Veenstra has designed a zero-emission fishing trawler featuring the use of alternative fuels, such as liquified natural gas (LNG) or electric propulsion. There are plans to construct the vessel out of fully recyclable material and for the full automation of the onboard fish processing.

Electric propulsion is arriving with aquaculture boats. *Astrid Helen* is a fully electric salmon farm workboat designed and built in 2019 by Grovfjord Mekaniske Verksted, Norway. She will save the planet from up to 91 tonnes (90t) CO2 and 900kg (1984 lb) of NOx particles annually – the average emission of a diesel powered fish-farm workboat. Fully electric work boats such as *Astrid Helene* will play a key role in the industry's future. Electric work boats are perfect for fish farming. The lack of engine noise is not only an advantage for the crew, but also for the salmon. It also reduces stress levels in the fish, and the environmental benefits are obvious.

Finally, the Toyota Motor Corp. is assisting the Japan Fisheries Research and Education Agency (FRA) with its plans to develop a fishing boat powered by hydrogen fuel cells.The first vessel will be a small boat to be used for tuna farming and is expected to go into commercialisation in 2022. It will work from a small, rural island called Goto, which has a lot of problems including diminishing population, high living and energy costs, and limited income sources.The island built a wind farm power plant at sea, but cannot use the produced electricity by itself. Then it also built a facility to produce hydrogen by use of the electricity, but no significant users exist in the island.If the local businesses such as fishing and fish farming can use this locally produced energy, the island does not have to buy high-priced oil, but can benefit from its own energy source.

Like agricultural drones, their equivalents, both aerially and underwater, are being employed for fish farming through monitoringwater conditions such as dissolved oxygen levels and using cameras to spot tears in nets. They can do this far more quickly than a boat or a diver.

In 2009, Jules Jaffe, biological oceanographer at Scripps Oceanography, UC San Diego created low-cost "miniature autonomous underwater explorers" or M-AUEs. The M-AUEs can be deployed in swarms to track planktonic movements as well as small-scale circulation patterns, oil spill dispersion or even sewage movement

Dr Tyler MacCready founded Apium Swarm Robotics, a Los Angeles based company to develop a low-cost marine drone for the purposes of monitoring and addressing chemical imbalances in the coastal waters of California, which led to his passionate pursuit of a scalable swarming solution. Swarms of drones can map ocean currents, temperatures, dissolved oxygen and salinity, all useful information for fish farmers. The drones can even monitor conditions in a fish pen while the fish are in it feeding.

Blue Robotics, founded in 2014by Erik Dyrkoren and Martin Ludvigsen at the Norwegian University of Science and Technology in Trondheim NTNU, manufactures a range of underwater drones such as the *BlueROV2* which gives fish farmers the opportunity to have a closer look at their fish welfare, how they behave and to do checkups on a regular basis. It is designed for optimal performance in all conditions from the Arctic oceans to tropical waters, and all the way down to 150 m (almost 500 ft) below the surface.

Japan's biggest tech and telecom companies, including Sharp, KDDI and NEC have teamed up with NTT Docomo and the University of Tokyo to test new technologies for oyster farming. The experiments are slated to run through March, 2021 in Hiroshima Prefecture, Japan's main oyster-producing region. Sensors attached to buoys and rafts measure water temperature and salt concentration, while drones search for oyster larvae and observe tides. This data is then analyzed to determine the best areas and time for attaching oyster spats to shells.

In August, 2015 members of the Hakai Institute carried out an aerial drone survey of eelgrass and kelp seaweed over the central coastline of British Columbia, Canada. They captured 415 images at an altitude of 300 metres (984 ft) over an area of approximately 4.5 km^2 (1.7 mi^2). In 2016, a unique collaboration between the Nature Conservancy, the University of California Santa Cruz (UCSC) and Hog Island Oyster Company in Tomales Bay, California aimed to fillanimportant knowledgegap: what wastherelationshipbetweenoyster aquaculture activity and extentand health of sensitive eelgrass habitats? To do this, they used a drone to monitor it from above.

It is likely that fish farming monitoring and maintenance will involve marine and aerial drones, electric marine agricultural machines working in synergy. Researchers from SINTEF and NTNU, together with developers from companies including Maritime Robotics, Argus Remote Systems and Lerow are currently working together to enable robotic systems to implement tasks currently carried out by people. The project, launched in 2016, is called ARTIFEX. The technology will be tested at SINTEF ACE's full-scale laboratory on the island of Frøya. The plan is to get an autonomous vessel to transport an ROV and a drone out to the facility.[3]

Endnotes

[1]Dary, op.cit.

[2]Kevin Desmond, Electric Boats and Ships: A History (Jefferson, NC.: McFarland, 2017).

[3]"Tomorrow 's fish farms will be unmanned" Feedstuffs Aug 28, 2017.

Appendix G

Electroculture

Alongside the experiments carried out into electroculture by von Maimbray, Bertholon, Siemens and others as mentioned in Chapter One, many scientists continued to work on the potential of increasing crop growth by electricity.

For example, in July, 1844 W. Ross reported to the Farmers' Club in New York that having planted potatoes in drills, he buried at one end of these rows a copper plate 1.5 mand 0.4 m deep (5 ft x 14 in), connected by a wire with a zincplate, the same size, buried two hundred feet away at the other end of the rows. On the 2nd of July, some of these potatoes were found 6.4 cm (2.5 in) in diameter, while the rows on either side, not under electric influence, had not formed tubers of more than 1 cm (0.5 in) indiameter. This statement was widely reported in the agricultural press[1]. With the massive emergence of chemical fertilizers at the end of the 19th century, many researchers attempted to demonstrate the possibility of improving yields by means of simple apparatus capable of increasing the receptivity and especially the adequate distribution of energy flows. In 1891, Frère Paulin, of the Agricultural School at Beauvais, France began a series of experiments near Montbrison, aimed at drawing down the atmospheric electricity in more abundant quantities than had been done by previous experimenters. He erected what he called a geomagnetifer, invented in 1843 by Berkensteiner. It was merely a tall, resinous pole planted in the earth, and carrying to its top a galvanized iron rod, insulated from it by porcelain knobs and terminating in five pointed branches. The electricity thus collected from the atmosphere was carried to the soil and distributed by means ofa system of underground wires to the area of ground to be electrically influenced. Frère Paulin's first experiment was with potatoes. The part of the field under electric influence responded in a surprising manner, as is evidenced by the following extract from a newspaper report at the time:

> The eye is arrested by a perceptible irregularity in the vegetation of the field. Within a circle limited exactly by the place occupied in the earth by the conducting wires of atmospheric electricity, the potato plants possess a vigour double that of the plants occupying the rest of the earth, and that, too, without a gap, without a

> feeble point in this group of superb stalks, sharply circumscribed as by a line drawn by a compass.

According to the report of the committee delegated by the Montbrison Society of Agriculture to report on Paulin's experiments, a geomagnetifer 8.5 m (28 ft) high, made its influence felt over a radius of 20 m (65 ft), and the yield of potatoes within this electrified area was from fifty per cent to seventy-five per cent greater than without it. This committee was quite enthusiastic over the results of Paulin's experiments, and awarded him a special medal. He next experimented with his geomagnetifer in a vineyard, and found that grapes were much advanced in their growth and that they were sweeter (yielding about five per cent, more sugar) and less acid. In further experiments he found that spinach and celery were markedly influenced, some leaves of the former reaching the length of 38 cm (1.25 ft) and some stalks of the latter 90 cm (3 ft). Radishes and turnips were much improved in size and quality, and sugar-beets yielded a larger percentage of their saccharine compound. It was also noticed that potatoes and sugar-beets electrically cultivated were singularly free from disease, while those outside of the influence were often seriously affected.[2] In 1896,Frère Paulin fitted a Poncelet wheel across the local river Thérain and installed an electric generator at Miauroy mill to power the "Ferme du Bois", fifteen years before neighbouring town of Beauvois was equipped with electricity.

As chronicled in Chapter One, the 1890s saw a craze for electro-ploughing, accompanied by the prevailing view that electroculture would produce giant fruits and flowers in any season but also destroy all insects that attack plants. In his novel "L'Île à hélice" ("Propeller Island), published in 1895, French science-fiction pioneer Jules Verne imagines a floating city, Milliard City, surrounded by artificial countryside and vegetation based on electroculture. During the first international meteorological conference held in Paris in the following year, it was decided to create a permanent committee on terrestrial magnetism and atmospheric electricity. In 1918, Raoul Marquis aka Henry de Graffigny, in his journal "Eureka" wrote about using an "Electrifier" for Intensive Cultivation by Electricity,imagining a town in Mesopotamia using electroculture.

Karl Selim Lemström, Professor of Physics at the University of Helsinki had noticed that trees growing under the *aurora borealis* were actually growing more rapidly than the same plants in warmer climates. Aware of the previous work on electroculture, Lemström attributed this growth to the electrical field generated by the Aurora, and set up a number of experiments to test this hypothesis. The experiments involved stringing a network of positively charged wires above a crop, charged to varying potentials for different periods of time, and measuring the difference in the yields obtained compared to controls. It was the results of these experiments that were the subject of Lemström's book

"*Electricity in Agriculture and Horticulture*", published in English in 1904.

Lemström's results, and his "overhead discharge technique" piqued the interest of a number of British scientists. The botanist J. H Priestley, working in cooperation with the physicist J. E. Newman, achieved a 17% increase in the yield of their cucumbers with Lemström's technique. Sir Oliver Lodge the famous physicist, again working with Newman, designed an 8 hectare(19.7 acre) installation based on Lemström's design and achieved a 24-39% increase in wheat grain yield. Lodge and Newman also set up the Agricultural Electric Discharge Company, Ltd. to commercialize their system.This in turn resulted in Lord Ernle, the president of the Board of Agriculture, later the Ministry of Agriculture and Fisheries (MAF), to found the Electro-Culture Committee in 1918 to see if Electro-Culture could be employed on a larger scale.

A range of large-scale field trials, took place both at both Rothamsted Experimental Station and Lincoln, electro-cultivating a wide range of crops, from wheat and oats to potatoes and cabbages. Including field trials on oats at Rothamsted between 1915 and 1917, by 1920 the Committee had amassed quite a considerable body of data. They had run twelve experiments, eleven of which found positive increases in yield, and eight of these were between 30% and 50%. However, although the next two years saw concerted efforts to build on this excellent start, poor weather, a very wet season in 1920 and a very dry one in 1921, led to very disappointing results.In 1921, an "economic installation" was built which was designed to be viable for actual farming (the cables were much higher off the ground so as not interfere with farm machinery and workers) and also to give a realistic cost for installing the apparatus. After experiencing

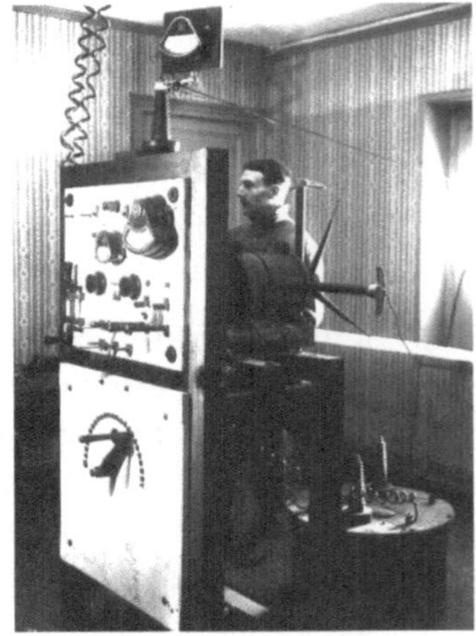

Electrification of plants at the National Office for Scientific and Industrial Research and Inventions at Meudon Bellevue, July 1920. (CNRS Photo Library / Historical Fund)

yet another disappointing season in 1922, the Committee decided to abandon the large-scale fieldwork.[3] Eventually, the Committee was axed in 1936 as a result of economic pressures, not because the idea was wrong.

Over in France, in 1923 independent researcher Justin Etienne Christofleau of La Queue-les-Yvelines, published "Augmentation des récoltes et sauvetage des arbresmalades per l'électroculture" and obtained patents concerning his Electro-MagnétiqueTerro Celeste. His system made use of "lightning rod" antenna, but with a buried antenna connected to buried north-south wires. Christofleau explained that it is not electricity as we know it but a breath of energy between heaven and earth, which stimulates and increases the fertility of the place. For the next twenty years, the Frenchman was persecuted for his inventions by lobbyists from the agrochemical industry who even attempted to have the word electroculture deleted from national dictionaries and encyclopaedias. In spite of this Christofleau's system was adoptedby farmers all over, in Australia, New Zealand, Africa and even China. He was not alone. In the August, 1935 issue of *Popular Science,* an article entitled "Electricity Controls Tree Growth" reported on the experiments of reputed French nurseryman Georges Truffaut at his laboratories in Versailles. He planned to invent the orchard of the future where it would be possible to control (advance or delay) the growth of trees and fruits.

ELECTRICITY CONTROLS TREE GROWTH

This drawing shows how electricity may be used to control an orchard of the future

ELECTRIFIED orchards are forecast by Georges Truffaut, French experimenter. Attaching wires from a forty-volt battery to seedling trees, he found their growth markedly stimulated when the current passed upward through the stems and branches. Reversing the flow retarded their development. Similar results could be obtained, Truffaut suggests, by fitting full-sized fruit trees with metal collars, connected to a suitable source of direct-current electricity. Thus a grower could retard the development of fruit to protect it against unseasonal frost, or hasten its ripening when conditions were favorable. To explain his observations, Truffaut offers the theory that the electric current alters the rate at which sap rises.

Tree at left shows result of stimulation by electric current

Downward flow of current retarded flowering of upper branches

Despite the domination of the pesticide and fertilizer industries, some have continued to practise electroculture. Yannick Van Doorne was born in 1976 in Ghent, Belgium. After obtaining his diploma in agriculture and biotechnology, set up a business called Ecosonic to research and develop the influence of electroculture, of music on plants, electromagnetic treatment of water, the "memory" and the reenergising of water. In 2008, Van Doorne started up Symphonie sarl. in Alsace to continue these developments. To produce giant unflowers, in 2009 Van Doorne used magnetic antennae made of beeswax.

Between 1976 and 2000, over 60 patents were taking out relating to electroculture (DC, AC, ES), magnetic and RF/microwave treatments to stimulate or retard plant growth. These included magnetic processes and equipment for treating and plantingseeds, a water-powered piezoelectric unit for producing nitrogen fertilizer.

Electroculture has finally been validated in China. Since the 1990s, Chinese scientists have been developing electroculture and in 2019, the Chinese Academy of Agricultural Sciences and other government research institutes released the findings of nearly three decades of study in areas with different climate, soil conditions and plantation habits. They are hailing the results as a breakthrough. Across the country, from Xinjiang's remote Gobi Desert to the developed coastal areas facing the Pacific Ocean, vegetable greenhouse farms with a combined area of more than 3,600 hectares (8,895 acres) have been taking part in an electroculture programme. The technique has boosted vegetable output by 20 to 30 per cent; pesticide use has decreased 70 to 100 per cent and fertiliser consumption has dropped more than 20 per cent. The vegetables grow under bare copper wires, set about 3 metres (10 ft) above ground level and stretching end to end under the greenhouse roof. The wires are capable of generating rapid, positive charges as high as 50,000 volts, or more than 400 times the standard residential voltage in the U.S. These high-voltage bursts kill bacteria and viral plant diseases both in the air and the soil. They also affect the surface tension of any water droplets on the leaves of plants, accelerating vaporization.

But can the electricity for electroculture come from sustainable sources?

In 1985, Jean-Marie Hangarter took out patent DE3665973D1 for an electroculture device comprising at least two grates or panels arranged parallel in the ground in a North-South direction. A voltage source provided an antenna comprising an electric current naturally occurring metal grids. It included the voltage source which consisted, firstly, of the sensing antenna of atmospheric electricity and, secondly, of one or several photovoltaic cells.

In 2019, Elenor M. Reyes, GlennJordan and M. Achico, researchers at *Batangas State University*, College of Engineering, in the Philippines proposed a *solar-*

powered electroculture technique for backyard farming. A portable solar power supply may be used to power up the project for a more cost-effective operation.

Finally, aged 72, Jimmy Luther Lee of Zenion Industries Inc., Rohnert Park, California, who has spent a lifetime inventing many electronic products which have been licensed to other companies in the fields of medicine, air cleaning and industrial controls, has come up with *Ion-A-Gro*, a solar-powered plant ioniser for enhancing plant metabolism (US2015070812).The apparatus includes a sealed and watertight housing, a solar cell array disposed atop the housing, and circuitry inside the housing for converting current generated by the solar cells into a pulsed high voltage discharged through an ion emitter.The circuitry includes a voltage regulator/converter circuit, an oscillator/modulator, a high voltage converter/multiplier, an ion emitter array, a photocell for switching the device off at night or in low light conditions, a light-emitting status indicator, an alternating current to direct current (AC/DC) wall adapter, and a hanger. Electroculture thus becomes more mobile.

Ironically, such devices use the natural outside light of our sun to shine on solar panels which in turn give power to electroculture which produces better vegetables but indoors. As electrical agricultural machines, they will play an integral part in the matrix of precision farming which includes satellites, drones, bots and micro-dripping, which may help resolve the challenging shortage of food in the years ahead.

Full circle?

In 2016, Tim Peake, the first British astronaut to visit the ISS, asked school children to help him with one of his scientific experiments. He wanted pupils to plant rocket seeds that have been in orbit with him and compare their growth with rocket plants that had stayed on Earth. While orbiting and monitoring the experiment, Peake would look out of a porthole of the ISS to catch sight of a large asterism consisting of seven bright stars of the constellation Ursa Major known to his ancient British ancestors as "The Plough". The plough which from horse and ox, to steam power, to electric cable systems, to petrol and diesel power and now back to electricity still travels up and down the starlit fields.

Endnotes

[1]"Galvanic Experiments on Vegetation", US Patent Office Report, 1844: pp 370-373.

[2]George S. Hull, "Electro-horticulture", New Rochelle, N.Y. The Knickerbocker Press, 1898.

[3]Snell, J., A.F. Berry, V. et al, "Interim Reports of the Electro-Culture Committee, 1919-1933.

In 2005, Kristian P. Olesen, a veteran in the HVAC industry, based in Stavengar, Norway developed a mixture of clay and water called Liquid Nano Clay. Sprinklers are used to spray the LNC half a meter into the sand, enabling the sand to hold water, and crops to be grown. The inorganic binder composition displays static electric charge, more precisely a homogenised dispersion of clay particle consisting substantially of single flakes of clay and air bubbles dispersed in a fluid. One major use of the binder composition is to reclaim arid and hyper-arid deserts and to prevent desertification and the movement and advancement of sand dunes, in other words stopping wind erosion efficiently. With this process any poor-quality sandy soil can be transformed into high-yield agricultural land in only seven hours. Olesen's vision and that of Desert Control, the firm he founded with Atle Idlund is to "Make Earth Green Again".

In a field test made using the world-patented LNC in the United Arab Emirates, two areas were planted with a selection of crops: tomatoes, aubergines and okra. One was treated with LNC while a second control area was left untreated. While the untreated area used almost 137 cubic metres of water for irrigation, the one treated with LNC used just 81 cubic metres, an enable saving of up to 52% of irrigation water and increase yields with less strain on scarce resources. Using LNC, deserts have been planted with over ten thousand trees, wheatfields, pepper fields. Other successful tests took place in Pakistan and China.

Olesen proposes that the biomass produced from desert-grown plants could provide clean electricity to power the desalinisation plants from which water could be used to irrigate the green deserts. Among the awards received by Olesen and his son is the World Wildlife Foundation's acclamation of Desert Control as a Climate Solver

Kristian P. Olesen "Inorganic, static electric binder composition, use thereof and method for the preparation of said binder composition" NO20060203 2006.

Bibliography

_______, *Electric Airplanes and Drones: A History*. Jefferson, NC: McFarland, 2018

_______, *Electric Boats and Ships: A History,* Jefferson, NC.: McFarland, 2017.

_______, *Electric Trucks: A History of Delivery Vehicles, Semis, Forklifts and Others*. Jefferson, NC: McFarland, 2019.

_______, *Innovators in Battery Technology: Profiles of 95 Influential Electrochemists*. Jefferson, NC: McFarland, 2015.

de Graffigny, Henry, *Electricité pour Tous*. Paris E Bernard, Imprimeur-Editeur, 1905. Chapter XV « Applications for Agriculture »

Desmond, Kevin, *Gustave Trouvé, French Electrical Genius (1839-1902),* Jefferson, NC: McFarland, 2015

du Moncel, Comte Theodose and Geraldy Frank. *L'Electricité comme Force Motrice*. Paris : Hachette,1884. (Deuxième édition) Collection « Bibliothèque des merveilles ».

Hull, George S, *Electro-horticulture*. New Rochelle, N.Y. The Knickerbocker Press,1898.

Rolt, L.T.C., *Great Engineers*, G. Bell and Sons Ltd, 1962.

Spence, Clark C, *Early Uses of Electricity in American Agriculture*. The Johns Hopkins University Press and the Society for the History of Technology, Vol. III 1962.

Index

A

B

C

D

E

F

G

H

N

O

P

Q

R

S